P9-DNC-266

Sugarless – Towards the Year 2000

Sugarless – Towards the Year 2000

Edited by

Andrew J. Rugg-Gunn

Department of Child Dental Health
University of Newcastle upon Tyne, UK

The Proceedings of an International Symposium 'Sugarless – Towards the Year 2000', held on 15–17 September 1993 at the University of Newcastle upon Tyne, UK

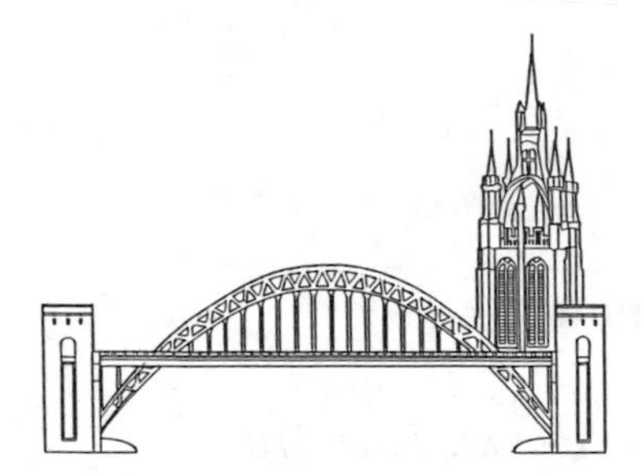

Special Publication No. 144

ISBN 0-85186-495-3

A catalogue record for this book is available from the British Library

Published by The Royal Society of Chemistry,
Thomas Graham House, The Science Park, Milton Road,
Cambridge CB4 4WF, UK

Printed in Great Britain by Bookcraft (Bath) Ltd

Preface

On 19-21 September 1990, an international symposium was held in the University of Newcastle upon Tyne, entitled "Sugarless - the Way Forward". This symposium, which was attended by 130 people from 9 countries, took place less than a year after the publication of the Department of Health's report[1] on "Dietary Sugars and Human Disease". This report, by the Committee on Medical Aspects of Food Policy, stated quite clearly that sugar consumption in the U.K. was too high and should be reduced. The symposium "Sugarless - the Way Forward" provided an opportunity to explore ways in which these reductions in sugar consumption could be achieved and the proceedings of the symposium were published in 1991[2].

Much has happened since 1990, so that a second "Sugarless" symposium was held in the University of Newcastle upon Tyne in September 1993 and the proceedings of this symposium are contained in this book. The most important event in the intervening three years was the publication of the Department of Health's report[3] on "Dietary Reference Values for Food Energy and Nutrients for the United Kingdom". For the first time in this country, quantitative limits were set by Government for consumption of non-milk extrinsic sugars by the population: these sugars should provide no more than 11% of food energy intake. A survey[4] of the nutrient intake of English adolescents in 1990, showed that 17% of food energy intake came from non-milk extrinsic sugars, indicating how far sugar consumption has to fall to meet the nutritional target for this group of the population.

Since the publication of the report[3] on dietary reference values by the Committee on Medical Aspects of Food Policy, a number of groups have been active in translating nutritional requirements into practical advice. One example is the report[5] of an expert working group on "Nutritional Guidelines for School Meals", published in 1992 by The Caroline Walker Trust. The Chair of that working group was Dr. Maggie Sanderson, whose paper "A dietitian's view of a sugarless diet", represented at the "Sugarless" symposium, is published in these proceedings.

Another event, which just preceded this Sugarless symposium, was the launch by the British Dental Association in July 1993 of the British Association for Toothfriendly Sweets. BATS, as it is known, is a non-profit-making organisation aimed at informing the public of the availability and benefit of toothfriendly non-sugar confectionery. Already in Switzerland, one-fifth of confectionery is sugar-free and "safe for teeth". The concept of Toothfriendly, as a way of encouraging progress towards a sugarless diet, was described by Dr. Albert Bär in the keynote lecture at the 1993 Sugarless symposium and his paper is the first in these proceedings. As Dr. Bär points out, the Toothfriendly idea has now spread to Germany, Belgium, France, U.K. and Japan.

Reviewing progress towards reducing sugar intake was an important aim of this symposium but it also provided a valuable opportunity for people from different backgrounds to meet and exchange views. Thus, the programme consisted of talks by nutritionists, dentists, food scientists, industrialists and experts on food legislation. The two principal groups of sugar-containing products targeted were confectionery and liquid oral medicines - similar to the emphasis given in the 1990 symposium. The papers included in these proceedings record the considerable progress that has occurred during the previous three years, as well as indicating areas where further work is required in order to facilitate and encourage the move towards a sugarless diet.

References

1. Department of Health, 'Dietary sugars and human disease', Report on health and social subjects 37, H.M.S.O., London, 1989.

2. A.J. Rugg-Gunn, 'Sugarless - the way forward', Elsevier Applied Science, London, 1991.

3. Department of Health, 'Dietary reference values for food energy and nutrients for the United Kingdom', Report on health and social subjects 41, H.M.S.O., London, 1991.

4. A.J. Rugg-Gunn, A.J. Adamson, D.R. Appleton, T.J. Butler and A.F. Hackett, J Hum Nutr Dietet, 1993, 6, 419.

5. Report of an expert working group, 'Nutritional guidelines for school meals', The Caroline Walker Trust, London, 1992.

A.J. Rugg-Gunn

Contents

Contributors

Dr. A. Bär
Toothfriendly Sweets International, Hauptstrasse 63, CH-4102 Binningen, Switzerland.

Miss D.W. Flowerdew
Department of Food Legislation, Leatherhead Food Research Association, Randalls Road, Leatherhead, Surrey KT22 7RY, United Kingdom.

Dr. A.J. Adamson
Enfield Community Care Trust, St. Ann's Hospital, St. Ann's Road, London N15 3TH, United Kingdom.

Dr. A. Maguire
Department of Child Dental Health, Dental School, Framlington Place, Newcastle upon Tyne NE2 4BW, United Kingdom.

Miss S.L. Johnson
Department of Food Legislation, Leatherhead Food Research Association, Randalls Road, Leatherhead, Surrey KT22 7RY, United Kingdom.

Professor A.S. Blinkhorn
Department of Oral Health and Development, University Dental Hospital, Higher Cambridge Street, Manchester M15 6FH, United Kingdom.

Dr. A.W.G. Walls
Department of Restorative Dentistry, Dental School, Framlington Place, Newcastle upon Tyne NE2 4BW, United Kingdom.

Dr. M. Sanderson
Division of Applied Chemistry, Life Sciences and Polymer Technology, University of North London, 166-220 Holloway Road, London N7 8DB, United Kingdom.

Dr. I.C. Mackie
Department of Oral Health and Development, University Dental Hospital, Higher Cambridge Street, Manchester M15 6FH, United Kingdom.

Dr. P.J. Sicard
Roquette Freres, F-62136 Lestrem, France.

Dr. Y. Le Bot
Roquette Freres, F-62136 Lestrem, France.

Mr. D.C. Pike
Vivil U.K. Ltd., Unit C, Bandet Way, Thame, Oxon OX9 3SJ, United Kingdom.

Professor A. Zumbé
Department of Biological Sciences, Salford University, Salford M5 4WT, United Kingdom.

Dr. A. Lee
Department of Biological Sciences, Salford University, Salford M5 4WT, United Kingdom.

Dr. D.M. Storey
Department of Biological Sciences, Salford University, Salford M5 4WT, United Kingdom.

Mr. K. Sugden
Reckitt & Colman Products, Dansom Lane, Kingston upon Hull HU8 7DS, United Kingdom.

Dr. I.G. Jolliffe
Reckitt & Colman Products, Dansom Lane, Kingston upon Hull HU8 7DS, United Kingdom.

Mr. A. Piotrowski
The Wrigley Company Ltd., Estover, Plymouth, Devon PL6 7PR, United Kingdom.

'Toothfriendly': Achievements after 10 Years and Future Prospects

A. Bär

TOOTHFRIENDLY SWEETS INTERNATIONAL, HAUPTSTRASSE 63, CH-4102 BINNINGEN, SWITZERLAND

1 INTRODUCTION

It is scientifically well established that frequent consumption of sugar, particularly between meals, increases the risk of caries formation [1,2]. The fact that caries is not yet an extinct disease demonstrates that ordinary oral hygiene and the use of fluoride as currently practiced cannot completely protect from dietary risk factors of caries [3,4]. In recognition of these circumstances, the World Health Organisation (WHO) and other authoritative bodies have therefore recommended to reduce the amount and frequency of consumption of sugars and/or to replace sugar by noncariogenic sweeteners in frequently consumed snacks [5-8].

Although most consumers have a theoretical knowledge about the relation between sugar intake and caries, the consumption of sugary confectionery is as widespread as ever, especially among children whose newly erupted teeth are particularly prone to caries formation and who often lack the skills and motivation to brush their teeth properly. Advice from dentists for many years to cut down on the intake of sweets has not had any measurable success and has probably not been taken very seriously by most of their patients. The reasons for the failure of such dietary counselling may be manifold but it would appear that for most people the emotional (subjectively positive) desire for sweets just overwhelms the rational (subjectively negative) knowledge about the risks of tooth decay.

In recognition of the importance of good dietary habits for dental health, and in recognition of the inappropriate dietary practices of many consumers, the Swiss University dental schools decided in the early 80s to change strategy and to launch a completely new public information campaign on nutrition and dental health. Instead of <u>indiscriminately</u> advising <u>against</u> the consumption of <u>any</u> sweets, the new campaign took a different, positive approach. The basic idea was that consumers should no longer be advised to avoid sweets altogether but that they should be educated and encouraged to eat only those sweet candies and gums that would not harm their teeth.

However, the problem with this approach was that consumers could not easily distinguish between dentally safe and potentially harmful sweets. Moreover, only a small number of sugarfree candies, mints and chewing gums were at that time available on the Swiss market and the variety of such products in terms of types, flavours, shapes and colours was too limited to give the consumer an attractive choice. Finally, there seemed to exist a wrong perception that sugarfree confectionery may only serve a useful purpose for weight control in adults but that it lacks otherwise any relevant health benefit, for example in relation to children's dental health.

Figure 1: Toothfriendly logo

The Swiss dental profession in collaboration with a few health-conscious confectionery manufacturers was therefore seeking ways to change this situation. The breakthrough was made in 1982 with the introduction of the "happy tooth" logo (Fig. 1). This eye-catching pictogram could be applied on the label of dentally safe toothfriendly confectionery in order to facilitate recognition of dentally safe products by the consumer. In parallel, the meaning of the logo was explained to the consumer in a general way by a non-profit association which was established for this purpose by the

Swiss university dental institute in cooperation with interested industry [9]. Since its foundation this association, known as "Aktion Zahnfreundlich", organises from time to time press conferences, maintains contacts to the media, and provides information on diet and oral health to school children through the school dental service and to adult consumers by means of directly distributed information leaflets. The impact of these generic public information campaigns is enhanced by product specific advertisements which are launched by the respective confectionery manufacturers and in which frequent reference to the "happy tooth" logo is made as well.

The results of these continued public information campaigns are impressive. By now (1993), 83% of Swiss consumers recognise the "happy tooth" pictogram and understand its significance for their personal oral health [10]. The consumer demand for toothfriendly sweets that arose from such knowledge pushed the market share of toothfriendly confectionery from about 6% in 1982 to about 20% in 1991/1992. In 1992, the Swiss consumer could choose between almost 20 different brands of toothfriendly chewing gum and about 60 different brands of toothfriendly candies. Taking also the various flavours of each brand into account, about 150 different toothfriendly products are now available to the Swiss consumers, including chocolate, chewable vitamin tablets, and an unsweetened, instant tea for infants.

In the present article, the reasons for this success and some basic facts about the "toothfriendly" public information campaign are outlined. It is hoped that the Swiss example will encourage the dental profession and the confectionery industry of other countries to launch identical dental health promotion campaigns in their countries. That this is possible is shown by the example of several countries in which "toothfriendly" campaigns have been implemented more recently.

2 THE "HAPPY TOOTH" PICTOGRAM

In our daily life we encounter numerous pictograms such as traffic signs, pictograms for "not smoking", symbols for men and women restrooms, etc. All these signs have instant recognition and mean the same thing to all of us. Most of them are understood even internationally, and by adults and children alike.

When in 1982 the Swiss university dental schools decided to draw the consumer's attention to the dental advantages of "Toothfriendly" products, they quickly realised that this could best be achieved by means of a pictogram which would signify "toothfriendly" properties. From a competition between different PR agencies, the "happy tooth" shown in Figure 1 emerged as the clear winner.

In order to keep control over the use of the pictogram, it was registered as a trademark first in Switzerland and subsequently in many other countries. Confectionery manufacturers who would like to use the logo on their labels and in advertising their products, can get the rights for the mark under a license agreement concluded with a national or the international Toothfriendly association. According to this agreement, the logo may be exhibited only on confectionery that does not promote tooth decay, i.e. on products that are friendly to the teeth (see next chapter for precise criteria to be fulfilled). The agreement also obliges the user of the mark to pay a small licence fee. This licence fee will be used by the Toothfriendly association for public information campaigns in the country in which the respective products are sold.

Ten years after its introduction, the eye-catching "happy tooth" may be found on toothfriendly confectionery sold in more than 20 different countries. The pictogram is easy to understand, even by children who cannot yet read and interpret food labels. It has an instant recognition and aids the health-conscious consumer to make the right choice. The "happy tooth" logo is therefore the centre-piece of the "Toothfriendly" public information campaigns.

3 HOW IS CONFECTIONERY TESTED FOR "TOOTHFRIENDLINESS"?

Whether a food is "toothfriendly" or not can be determined by means of an in vivo plaque-pH-telemetry test. In Switzerland, this test method has been used for more than 20 years. On an international level, the validity of the method was recognised in 1985 when more than 70 scientists from 8 countries reached a consensus on the methods to be used for assessing the cariogenic potential of food. It was concluded that appropriately conducted plaque-pH-telemetry tests are sufficient to identify non-acidogenic or hypo-acidogenic foods, i.e. foods that even under conditions of frequent consumption do not present a significant caries risk for human teeth [11].

Detailed descriptions of the plaque-pH-telemetry test have been published in the scientific literature [12,13]. In the present context, it may suffice to say that this test measures the formation of acid by plaque bacteria from ingested food in vivo. Volunteers who carry a partial prosthesis with an implanted pH electrode, consume the test food in the ordinary way. The oral bacteria that during a preceding 3- to 4-day period have formed a plaque over the electrode, come thereby into contact with the ingested food. If the food contains fermentable components, such as sugars and starch, acid will be formed. On the other hand, non-fermentable ingredients, such as sugar alcohols (polyols), will not result in significant acid formation. The concentration of the acid formed is measured by the implanted electrode under the plaque, i.e. at the site at which acid could lead to demineralisation of the tooth enamel and consequently to caries. If the measured acidity does not exceed a certain, critical limit (pH 5.7) during and for 30 minutes after consumption of the test food, the product is considered to be non- or hypo-acidogenic (Fig. 2). Products that comply with this criterion, and do not contain excessive amounts of acidulants that may cause erosion by direct contact with the tooth, are considered "toothfriendly" and qualify for distinction with the "happy tooth".

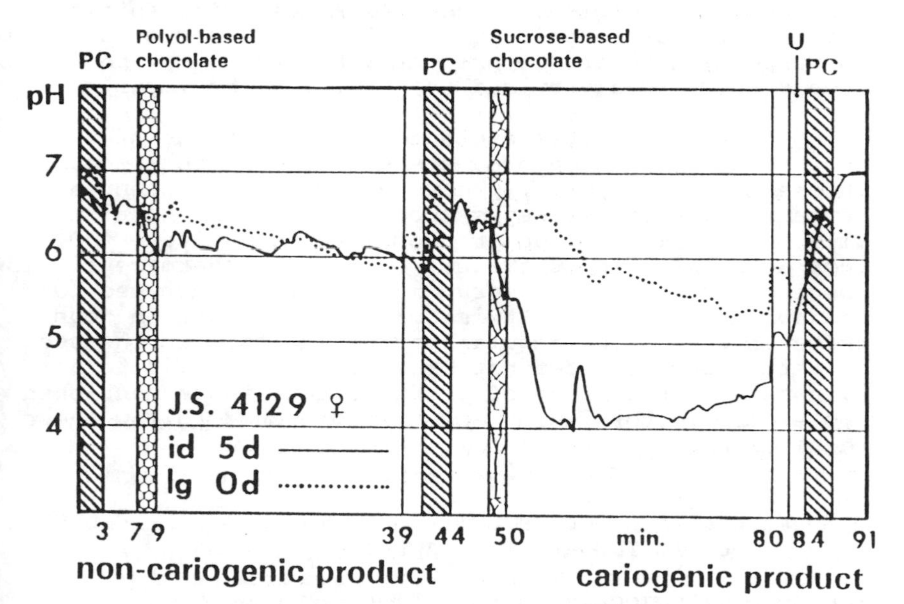

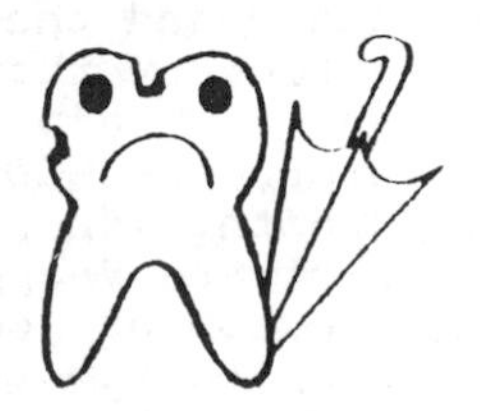

Figure 2: Telemetrically recorded pH after consumption of a non-cariogenic and a cariogenic product

With the introduction of an increasing number of toothfriendly confectionery products exhibiting the "happy tooth" prominently on the label, a number of producers of other products inquired whether they also could use the logo on their products such as toothpaste, sugarfree medications, bread spreads, and different types of beverages. In order to assure fair and equal treatment of the interested industries, and to protect the consumer from potentially misleading or confusing uses of the pictogram, the Toothfriendly associations limited the range of products on which the toothfriendly logo may be exhibited to:

(a) foods which, under normal conditions of use, lack the acidogenicity and thus the cariogenic potential that characterises the corresponding reference foods, and which are nutritionally not inferior to the foods they substitute for; and

(b) OTC medicinal preparations which, under the recommended conditions of use, lack the acidogenicity and thus the cariogenic potential that characterises the corresponding reference products, and which may be freely sold and advertised to the public at large.

These definitions excluded the use of the toothfriendly logo on products such as cheese (because cheese does not typically have a cariogenic potential), certain types of beverages (because they may nutritionally be inferior to the products they substitute for, such as milk or fruit juice), or ethical pharmaceuticals (because an antibiotic syrup may not be advertised to the public at large). It is self-evident that in countries in which food regulations would restrict the use of terms such as "does not promote tooth decay" to an even more narrow range of products than those mentioned above, the Toothfriendly association would follow these regulations.

4. WHAT ARE THE OBJECTIVES OF THE "TOOTHFRIENDLY" PUBLIC INFORMATION CAMPAIGNS ?

The general objective of the "Toothfriendly" public information campaigns is to improve dental health by promoting toothfriendly dietary habits. In particular, the campaigns intend to inform the consumer about the existence and the dental benefits of toothfriendly confectionery that may be consumed even frequently without any harm to the teeth. The campaigns implement therefore the WHO recommendation mentioned above [5]. In the "Toothfriendly" information campaigns the significance of the "happy tooth" and the relevance of the consumption of toothfriendly products for dental health is explained to the consumer in order to enable him to easily make the "healthier choice".

In the context of the toothfriendly campaigns, dialogue with confectionery manufacturers is also sought. Since adult consumers rarely read the fine-print of food labels, and since children are usually unable to read and understand such information, confectionery manufacturers realised that claims such as "sugarfree" do not always provide sufficient information to the consumer and that the "happy tooth" makes products that fulfill the "toothfriendly" quality criteria more attractive. Experience from the Swiss market demonstrates that in fact only the "happy tooth" enables adults and children to readily identify products that do not harm the teeth.

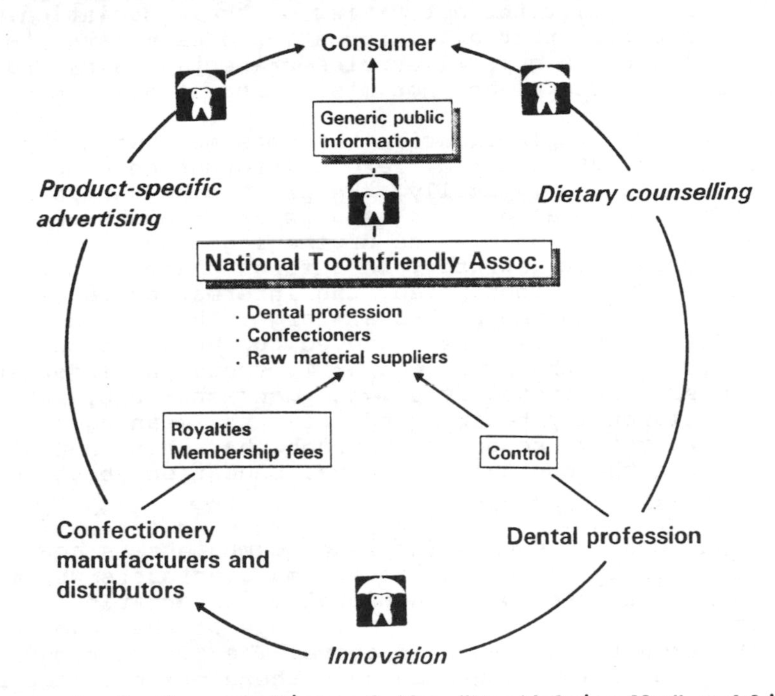

Figure 3: Implementation of the "Toothfriendly" public information campaigns.

5. HOW IS THE "TOOTHFRIENDLY" MESSAGE BROUGHT TO THE CONSUMER ?

Three parties participate in the "toothfriendly" caries prevention programme: the national (or international) toothfriendly associations, the practicing dentists and the confectionery industry (Fig. 3).

In order to coordinate the toothfriendly public information campaigns and in order to foster the contacts between confectionery manufacturers and the dental profession, non-profit Toothfriendly associations have been established in Switzerland (1982), Germany (1985), France (1990), Belgium (1992), the U.K. (1993) and, most recently, Japan (1993). Members of these national associations are representatives of the dental profession (university dental schools, national dental associations, etc.), manufacturers or distributors of toothfriendly confectionery, and suppliers of the raw materials used for their production. For implementation of the toothfriendly programme in countries in which a national association is not yet established, an international association ("Toothfriendly Sweets International") is responsible. The statutes of all Toothfriendly associations ensure that the dental profession has at any time full control over the activities of the association. This guarantees that all information disseminated to the public pursues a general, caries-preventive aim, and that it is truthful and not misleading at all times.

In order to familiarise the consumers with the "toothfriendly" label, different information campaigns are implemented. Typically, the practicing dentists and other professional opinion leaders are brought up-to-date about the significance of the logo and the new caries preventive strategy in a first phase of the campaign. In a second phase, the information campaign is then directed to the public at large. The "toothfriendly" message may be conveyed to the adult consumer by articles in the print media, TV-spots and information brochures distributed by private dentists or by retailers. Children may be informed about the meaning of the "happy tooth" at school or through the school dental service in the context of general education about dental health.

The cost of such information campaigns is covered by payments by the confectionery manufacturers who must pay a minimal fee to the national or international toothfriendly association for the use of the logo on their products as explained above. The "Toothfriendly" caries prevention programme does therefore not require funds from government-sponsored public health programmes.

5. WHAT RESULTS HAVE BEEN ACHIEVED WITH THE "TOOTHFRIENDLY" CAMPAIGN ?

In Switzerland, the toothfriendly logo was launched by the Swiss Toothfriendly Association ("Aktion Zahnfreundlich") a bit more than 10 years ago. At that time, the market share of sugarfree confectionery represented a meager 6%. However, five confectionery manufacturers recognised the opportunity that the programme and the active support by the dental profession could provide for their products. The caries-preventive hopes of the dental professionals and the commercial hopes of the confectionery manufacturers were fulfilled. The market share of toothfriendly confectionery rose from about 6% in 1982 to about 20% in 1990/92 [14]. This corresponds to a total volume of nearly 4000 tons for a total population of about 7 million inhabitants. Also in the most recent past, the growth of the sugarfree segment out-performed the traditional, sugar-based one (Fig. 4). At the end of 1992, over 30 confectionery manufacturers and distributors sold in Switzerland about 60 different brands of toothfriendly candies, nearly 20 different brands of toothfriendly chewing gums, and one toothfriendly chocolate (different flavours not counted as different products !). Since in other European countries the market of sugarfree confectionery was expanding much more slowly, the more rapid development of the Swiss market may be attributed at least partly to the various "toothfriendly" public information campaigns.

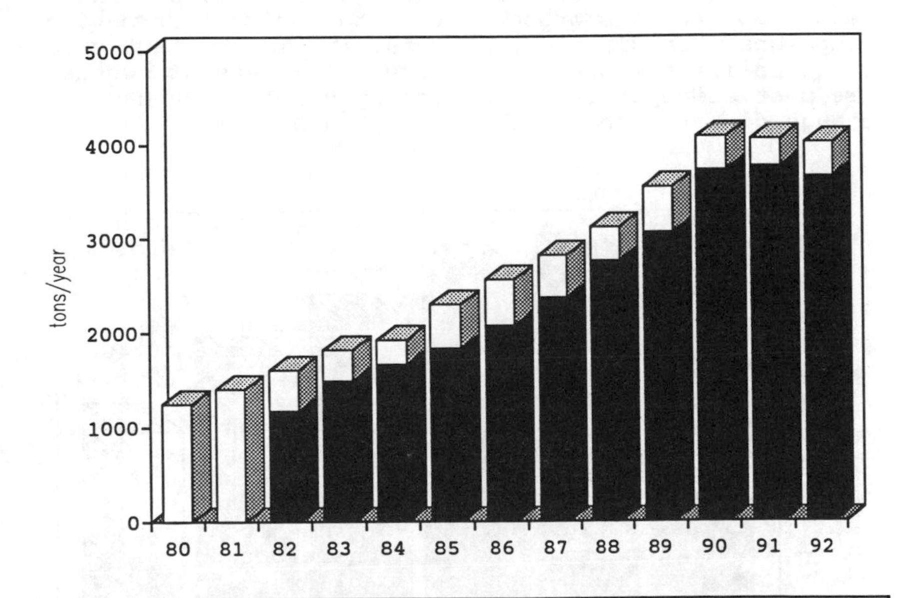

■ Sugar substituted products carrying the Toothfriendly pictogram

□ Sugar substituted confectionery products without the Toothfriendly pictogram

Note: The per capita consumption of confectionery is about 3 kg/year (1963: 2,7 kg; 1968: 2,8 kg; 1973: 3,2 kg; 1978: 2,9 kg; 1983: 2,9 kg; 1988: 3,1 kg; 1992: 3.0 kg)

"Confectionery" includes chewing gum, hard-boiled candies, toffees, nougat, marzipan, lozenges including medicated OTC products (cough lozenges), etc. It does not include chocolate or chocolate-coated products

Figure 4: Sales of sugar-substituted confectionery in Switzerland from 1980 to 1992

During the past ten years, the Swiss Toothfriendly Association launched articles and advertisements in the print media, distributed leaflets to opinion leaders (dentists, teachers, nutritionists, etc.), distributors (e.g. kiosk owners) and the consumers, and informed all school-age children with particular care about the meaning of the "happy tooth". The result of these and other activities in terms of consumer awareness and recognition of the logo is impressive. Four representative awareness analyses were conducted by an independent market research institute in 1986, 1989, 1991 and 1993, i.e. at 4, 7, 9 and 11 years after start of the programme [10,15,16]. It was found that at the first three occasions about three out of four consumers recognised the toothfriendly logo. In 1993, the awareness rose on an average of all age groups to 83%. Interestingly, the recognition of the logo was highest in school age children and decreased with increasing age (Fig. 5). The reason for this is probably that the Swiss information campaigns were directed, at least initially, particularly at children who after all are big lovers of sweets, are most susceptible to developing caries, and may change dietary habits more easily than adults.

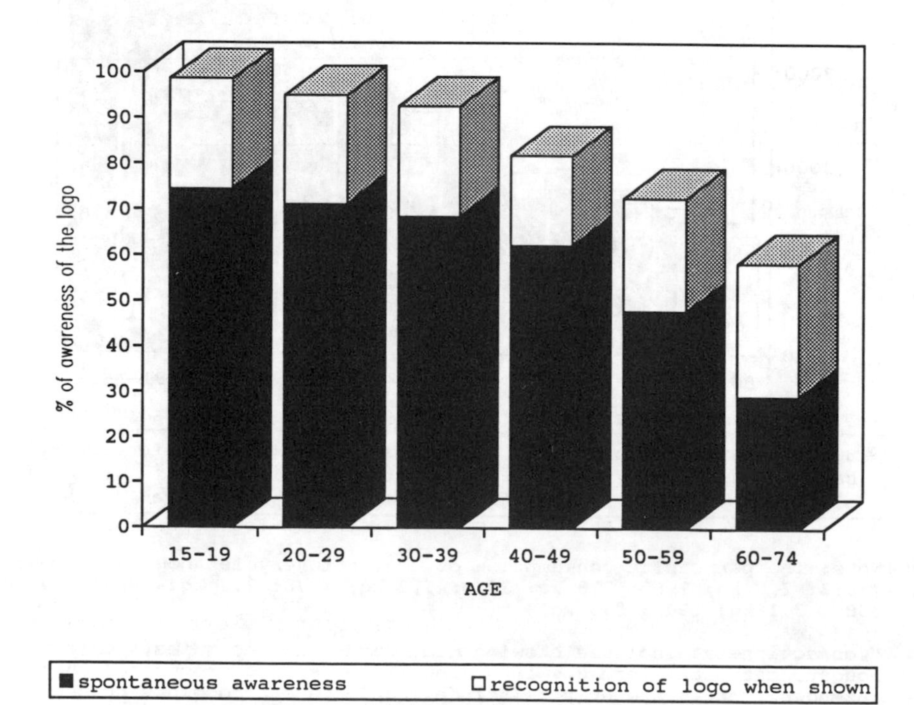

Figure 5: Recognition of the "happy tooth" logo in different age groups of Swiss consumers in 1993.

The Swiss awareness analyses revealed also that the motivation to buy toothfriendly products is remarkably high. On average, over the four analyses, about 20% of the persons who knew the "happy tooth" claimed to buy essentially only products with this pictogram. An additional 50% declared they buy such products at least from time to time. Among those who consume confectionery (gums, candies, etc.) daily, 42% claim to buy principally only products with the logo. However, the interest in the logo, not unexpectedly, decreases with decreasing frequency of confectionery consumption as well (Table 1). Among those persons who know the "happy tooth", the intention to buy toothfriendly products does not differ between the different age groups or socio-economic classes. However, women tend to buy toothfriendly products somewhat more often than men [10].

Table 1: Association between frequency of confectionery consumption and intention to choose products with the "happy tooth" (Swiss consumer survey 1993)

Intention to buy confectionery with the logo	Frequency of confectionery consumption			
	daily	a few times per week	a few times per month	almost never
always	42	18	14	17
from time to time	41	63	56	36
never	2	4	7	21
don't look for logo	14	14	22	24
don't know	0	0	0	2

Note: Figures denote percentage of consumers, values in a column add up to 100%.
For the survey, a representative sample of 1000 Swiss inhabitants (age 15-74) was interviewed. The figures presented in this table are based on the responses of those 830 persons who recognised the logo spontaneously or when shown.

The positive attitude towards toothfriendly products becomes apparent in the Swiss analyses also from the number of persons who would like to see the toothfriendly label on additional products. For the time being, the label appears mostly on sugarfree chewing gum and candies. Considering the existing number of toothfriendly candies and gums on the Swiss market, it is not surprising that the Swiss consumer is satisfied with the available variety and that relatively few (21%) would like to see more products with the toothfriendly label. Of these only 7% would like to see more toothfriendly chewing gum and 14% more candies. On the other hand, about 50% would like to see more toothfriendly chocolate, which indeed is not yet available on a large scale. These results illustrate that the consumers truly look for a larger variety of certain toothfriendly products [10].

The results of the Swiss public awareness analyses demonstrate convincingly that the labelling of toothfriendly products with the "happy tooth", in conjunction with a credible and well-targeted information campaign, results in a high public awareness and knowledge about toothfriendly dietary habits. The motivation to buy such products is correspondingly high. One may therefore be confident that the toothfriendly campaign will become a success also in those countries in which the programme has been launched only recently. The observation of an increase in the sales of toothfriendly products in Germany from about 6000 tons in 1991 to 8000 tons in 1992 supports this notion.

While changes in confectionery consumption habits can be quantified quite accurately, it is much more difficult to estimate the caries preventive effect of this change [17]. In Switzerland which would be best suited for a respective investigation since this country has the longest history of "toothfriendly", most consumers use such products at least from time to time [10]. It is therefore virtually impossible to identify an "untreated" control group which does not consume any such products but which for all other relevant parameters matches the characteristics of the "user" group. Moreover, a very large number of subjects would have to be examined since caries prevalence in Switzerland is very low. Estimates of the caries preventive effect of sugar substitution may therefore be derived only indirectly from studies with controlled administration of sugar-substituted sweets or from analysis on the relation between self-reported sugar intake and caries experience. The weight of evidence from such data indicates that sugar substitution helps prevent caries even in fluoridated populations [1,18-22]. Common sense would dictate that those groups of the population who exhibit the highest caries prevalence would benefit most from a change of dietary habits. In this context, it appears relevant that the "happy tooth" message may be understood by children, the less educated, or immigrants who represent segments of the population with a particular risk for developing caries.

6. CONCLUSIONS AND OUTLOOK

Frequent consumption of sugar(s) continues to be a risk factor for caries. If it is agreed that a preventable disease should be prevented, it is reasonable to substitute sugary sweets between meals by sugarfree alternatives that are safe for the teeth. By means of the "happy tooth" logo, different national toothfriendly associations try to bring this message to the consumer. The rising sales of toothfriendly products worldwide reflect the success of this approach and the increasing consumer demand for such products (Table 2).

Table 2: Sales of toothfriendly sweets in 1992

Countries	Sales 1992 in kg
European Community	10,055,709
Other European Countries	4,599,078
North, Middle and South America	253,145
Far East	226,019
Middle East	25,218
Others	2,628
TOTAL	5,161,797

The "toothfriendly" approach has been successful because the appealing logo ensures instant recognition, the information campaigns organised under the auspices of the dental profession have a high credibility and media attention, the toothfriendly products are nowadays as tasty as the traditional sugar-based ones, the recognition of the logo is growing automatically as the number of toothfriendly products increases and such products are "easy to get". The representatives of the dental profession have no hesitation to identify themselves with the objectives of the programme since they are not misused to endorse specific products or ingredients when in fact the substitution of sugary sweets by any toothfriendly alternative is to be advocated from a caries-preventive point of view.

Considering the still widespread consumption of sugar-based products between meals, and the above-average caries incidence in certain risk groups of the population, or in some countries in which caries is still high or even on the rise, it is the long-term goal of the toothfriendly associations to not only inform the consumers in their country about the advantages of toothfriendly dietary habits but to spread this message also internationally. The recent introduction of toothfriendly products in Turkey, Japan and South Korea represent the first results of this approach.

REFERENCES

1. A.J. Rugg-Gunn, 'Nutrition and Dental Health', Oxford University Press, Oxford, 1993, Chapter 6, p. 113.

2. E. Newbrun, in: 'Sugars and Sweeteners', N. Kretchmer & C.B. Hollenbeck (eds.), CRC Press, London, 1991, Chapter 10, p. 175.

3. W.H. Bowen, J. Am. Dent. Assoc., 1991, 122, 49.

4. I.D. Mandel, Int. Dent. J., 1993, 43, 67.

5. WHO, 'World Health Organization Technical Report Series', Geneva, 1984, no 713, p. 19.

6. Committee on Medical Aspects of Food Policy, Report no 37, 'Dietary Sugars and Human Disease', Her Majesty's Stationery Office, London, 1989, Chapter 6, p. 16.

7. The Surgeon General's Report on Nutrition and Health, US Department of Health and Human Services, Washington D.C., 1988, Chapter 8, p. 345.

8. National Research Council, 'Diet and Health: Implications for reducing chronic disease risk', Natl. Academy Press, Washington, 1989, p. 637.

9. T. Imfeld & B. Guggenheim, in: 'Sugarless - The Way Forward', A.J. Rugg-Gunn (ed.), Elsevier Applied Science, London, 1991, p. 197.

10. IHA-GFM, Sympadent-Zahnmännchen, unpublished report no 619.1519.51 prepared for Aktion Zahnfreundlich, Nov. 1993.

11. Working Group Consensus Report, J. Dent. Res., 1986, 65 (Spec. Issue), 1530.

12. T.N. Imfeld, Monographs in Oral Science, 1983, 11, 1.

13. D.S. Harper, D.C. Abelson & M.E. Jensen, J. Dent. Res., 1986, 65 (Spec. Issue), 1503.

14. A. Bär, in: 'Food Ingredients Europe Conference Proceedings', Expoconsult Publishers, The Netherlands, 1990, p. 12.

15. A. Bär, in: 'Food and Nutrition Policy', Second European Conference on Food and Nutrition Policy, The Hague, 1992, p. 88.

16. J.-C. Salamin, in: 'Sugarless - The Way Forward', A.J. Rugg-Gunn (ed.), Elsevier Applied Science, London, 1991, p. 85.

17. T.M. Marthaler, Caries Res., 1990, 24 (suppl. 1), 3.

18. B.A. Burt, Caries Res., 1993, 27 (suppl. 1), 56.

19. A. Bär, Lebensm.-Wiss. u.-Technol., 1989, 22, 46.

20. J. Klimek, P. Rauch, E. Hellwig & H. Prinz, J. Dent. Res., Abstract 112, 1988, 67, 696.

21. A.S. Papas, A. Joshi, A. Belanger & C. Palmer, Caries Res., 1992, 26, 221.

22. M.P. Faine, D. Allender, D. Baab, R. Persson & R.J. Lamont, Spec. Care Dentist, 1992, 12, 177.

Current Legal Status of Non-sugar Sweeteners in Europe

Dorothy W. Flowerdew

DEPARTMENT HEAD – FOOD LEGISLATION, LEATHERHEAD FOOD RESEARCH ASSOCIATION, RANDALLS ROAD, LEATHERHEAD, SURREY KT22 7RY, UK

1 INTRODUCTION

The comparatively quick development of laws that regulate the use of non-sugar sweeteners in foodstuffs is one of the most interesting aspects of the European Community food harmonisation programme as it stands today. Laws that authorise the use of sweeteners in foods have been in force in the United Kingdom and in other European countries for many years, and it is necessary to focus on these first in order that the EC position can be clearly understood.

2 UNITED KINGDOM LEGISLATION

For many years saccharin and its sodium and calcium salts were the only artificial sweeteners permitted for use in food. Cyclamate and its calcium and sodium salts were allowed for a short time, but doubts on their safety-in-use led to withdrawal from the permitted list by the Artificial Sweeteners in Food Regulations 1969[1]. Compositional regulations allowed saccharin to maximum levels in soft drinks and prohibited its use in ice cream and, later, in jam, except that it could be used in reduced-sugar and diabetic jams. Saccharin and its salts have not been restricted in other foods unless a particular standard prohibits the use of additives generally. However, the general provisions of the Food Safety Act 1990 (and its predecessor acts) prohibit the sale of food that is injurious to health, or which misleads the consumer as to its nature, substance and quality. These provisions have been backed up by recommendations regarding Good Manufacturing Practice, that state that additives should not be used in foods in quantities exceeding those needed to achieve the desired technological effect in the food. These general provisions, which are legally enforceable, have provided effective controls over the use of most classes of additives used in foods in the United Kingdom.

The Artificial Sweeteners in Food Regulations[1] include a definition of "'artificial sweetener': any chemical compound which is sweet to the taste but does not include any sugar or polyhydric alcohol". By reason of this exclusion, polyhydric alcohols could be used as sweeteners. Further controls were exerted when the Miscellaneous Additives in Food Regulations[2] came into force in 1980 because these regulations included sorbitol, sorbitol syrup and mannitol in a long list of additives authorised for various functions in foods. Similarly, glycerol has been permitted as a carrier solvent. The mechanism of UK legislation allows that these polyhydric alcohols, although not listed as sweeteners, may be used for that purpose, apart from any restrictions imposed by the Act or specific regulations made under it. Therefore saccharin and its sodium and calcium salts, sorbitol, sorbitol syrup, mannitol and glycerol could be used as sweeteners in foodstuffs.

In 1977 a full review of sweeteners for use in foods was undertaken by the Food Additives and Contaminants Committee (FACC) at the request of Ministers; the review was stimulated by doubts concerning the safety-in-use of saccharin and requests for newly developed sweeteners to be considered for authorisation for use in food and drink products. The important Report of the Committee was published in 1982[3]. The Committee reviewed 25 substances in all, both the 'intense' and the 'bulk' sweeteners, considering their technological need and also the toxicological recommendations of the Committee on Toxicity of Chemicals in Food, Consumer Products and the Environment (COT), whose Report appends the FACC Report. The Committee recommended that bulk and intense sweeteners should be controlled by a permitted list and that new regulations should be made to control their use. Subsequently, in 1983, the Sweeteners in Food Regulations were issued[4]. In addition to saccharin and its salts, six new sweeteners were authorised for use: the intense sweeteners acesulfam, aspartame and thaumatin; and the bulk sweeteners hydrogenated glucose syrup, isomalt and xylitol. Sorbitol and mannitol were removed from the control of the Miscellaneous Additives in Food Regulations and relisted in the Sweeteners in Food Regulations. Sweeteners, other than sugars, are prohibited in foods manufactured specially for babies and young children (except special dietary products). Existing regulations were also amended to allow the new sweeteners to be used in soft drinks, diabetic jams and the intense sweeteners in reduced-sugar jams. As before, they can also be used in foods generally provided there are no other restrictive legal provisions. The Sweeteners in Food Regulations are still in force, having been amended to add lactitol to the permitted list.

List of non-sugar sweeteners authorised for food use in the United Kingdom:-

acesulfam potassium
aspartame
hydrogenated glucose syrup
isomalt
lactitol
mannitol
saccharin
sodium saccharin
calcium saccharin
sorbitol
sorbitol syrup
thaumatin
xylitol
glycerol*

* permitted by the Solvents in Food Regulations

3 OTHER EUROPEAN COUNTRIES

By contrast with the United Kingdom approach, most European countries have much more restrictive laws that control the types of additives that may be used in foods, their levels of use and the foods that may contain additives. Additionally, in the case of non-sugar sweeteners, the concept of using these in special-purpose and dietary foods only is much more strongly held than in the UK. Consequently, in many cases sweeteners are restricted to use in a few foods that are modified for dietetic purposes - to aid slimming and weight control, to provide special foods for diabetics or to reduce the incidence of dental caries. A few examples follow.

Belgium

There is a permitted list of sweeteners (which includes neohesperidine dihydrochalcone), but detailed regulations are very restrictive and generally sweeteners are only allowed in certain foods for total replacement of sugars. For instance, acesulfam is permitted in reduced-carbohydrate soft drinks, maximum 600 mg/litre; aspartame in dietetic 'limonade', max. 500 mg/l, low-fat fermented milks, low-fat yoghurts and similar, max. 600 mg/kg, and confectionery gums, max. 4 g/kg; NHDC is permitted in chewing gum, max. 400 mg/kg and dietetic 'limonade', max. 50 mg/l; maltitol, sorbitol and xylitol are allowed in chewing gums and sweets to GMP, all authorisations for the total replacement of sugars.

France

There is a list of permitted sweeteners including acesulfam, aspartame, saccharin and its ammonium, calcium, potassium and sodium salts and the bulk sweeteners. Detailed authorisations encompass low-carbohydrate foods only, including acesulfam in low-carbohydrate dietetic non-alcoholic beverages based on fruit juices, max. 360

mg/l, calculated on the ready-to-drink product, low-carbohydrate dietetic chewing gum, max. 5000 mg/kg, and table-top sweeteners; aspartame in low-carbohydrate dietetic beverages based on fruit juices, max. 600 mg/l, calculated on the ready-to-drink product, low-carbohydrate chewing gum, max. 6000 ppm, low-carbohydrate dietetic edible ices, ice cream and sorbets, max. 1000 ppm, and low-calorie dietetic products, max. 100 mg/100 kcal product. Bulk sweeteners such as isomalt, lactitol, maltitol, hydrogenated glucose syrup and polydextrose are permitted in foods for particular uses with no upper limits of use set, and sorbitol is also allowed in low-carbohydrate products, max. 20%.

Germany

Authorisations are few. Cyclamates and saccharin are permitted in some dietetic foods. Saccharin is also allowed in 'Brause' (artificially sweetened soft drinks), max. 200 mg/l, chewing gum and wafers. Mannitol, sorbitol and xylitol are allowed as sugar substitutes in dietetic foods and in chewing gum.

The Netherlands

Sweeteners are permitted in certain low-energy soft drinks: acesulfam max. 600 mg/l, aspartame max. 750 mg/l, cyclamates max. 400 mg/l and saccharin max. 125 mg/l; they are also permitted in table-top sweeteners. Cyclamates and saccharin are allowed in diabetic ice desserts that are approved by the authorities. Sorbitol, mannitol and glycerol are allowed for use as sweeteners, and also function as emulsifiers and stabilisers.

4 THE EUROPEAN COMMUNITY

Up to 1985 harmonisation of food laws in the EC was directed towards agreement on laws that affected specific horizontal areas of food manufacture, such as quantity control and labelling, or specific commodities including fruit juices and jams. Four directives covering additives were agreed: antioxidants; colours; emulsifiers, stabilisers, thickeners and gelling agents; and preservatives. Authorised lists of additives were prescribed, and it was intended that detailed food uses would follow. However, with few exceptions such detailed second-order controls did not materialise, leaving Member States free to impose their own controls on the foods in which authorised additives could be used and to prescribe usage levels in whatever way they considered necessary. Harmonisation in its true sense did not really exist in the area of food additives, and manufacturers and suppliers of food additives who wished to export their foods/products to other EC countries have had to understand and comply with different restrictions in every country - a far cry from the intended Common Market!

The Single Market Initiative

Following the issue of the White Paper in 1985, the EC Commission re-oriented its strategy for harmonisation of food laws. A target date of 1 January 1993 was set, by which harmonisation of laws would be complete and a true 'Single Market' established. Rules were to be made in the interests of protection of health, provision of information to consumers, ensuring fair trading and provision for necessary public controls only. Since then work in many areas has been dropped or changed. However, since control of additives falls within the provision regarding protection of health, rules would be developed and include limits of use in prescribed foods, where necessary. In order to accomplish this enormous task by the target date of January 1993, the concept of a 'framework' directive was developed; this would set generally applicable rules and 'specific' directives setting detailed rules would be made under it. This approach resulted in acceptance of Directive 89/107/EEC of 21 December 1988 concerning food additives for use in foodstuffs[5], the framework additives directive. Its provisions include definition of 'additive', criteria for acceptance of food additives and procedures to be followed if a Member State wishes to prohibit an additive on health grounds or, in the interests of scientific or technological development, a Member State wishes to authorise provisionally a substance not yet included on a positive list. Most importantly, the framework directive requires that a comprehensive directive be developed; this may be drawn up in stages and will include an exclusive list of additives and a list of foods to which these additives may be added, with a limit on their use where necessary.

Proposed Draft Directive on Sweeteners for Use in Foods

The development of such detailed laws was spearheaded by a proposal to control the use of sweeteners in foods, the first stage of the required comprehensive directive. It was followed by proposed drafts on colours for use in foodstuffs, and a third draft to cover all other classes of additives, including preservatives, emulsifiers, flavour enhancers, etc. Considering the differences in existing national laws, the proposal on sweeteners was tabled fairly quickly. It was stated that the use of sweeteners to replace sugar is justified for the production of energy-reduced foods (by at least 30% compared with a similar unmodified product), non-cariogenic food-stuffs or foodstuffs without added sugar (meaning without added mono- or disaccharides or any food used for sweetening, except that diabetic foods may contain fructose), for the extension of shelf-life through the replacement of sugar and for the production of dietetic products. Thus the European concept of the use of sweeteners in special and dietetic foods, rather than the more liberal provisions of UK laws, prevailed during early discussions. Nevertheless, a number of foods more generally consumed in the diet have subsequently been included in the proposed list of foodstuffs to which sweeteners may be added.

The proposed directive applies to sweeteners used both to impart a sweet taste to foodstuffs and as tabletop sweeteners. It does not apply to foodstuffs used for their sweetening properties (such as honey and fruit juice). There is a general prohibition on the use of sweeteners in foods specially prepared for infants and young children. All the listed bulk and intense sweeteners have been evaluated by the EC Scientific Committee for Food, whose recommendations have been taken into account when setting limits for use in food and drink products. The following sweeteners are listed:

Bulk sweeteners:	E 420	Sorbitol, sorbitol syrup
	E 421	Mannitol
	E 953	Isomalt
	E 965	Maltitol, maltitol syrup
	E 966	Lactitol
	E 967	Xylitol

These may be used to <u>quantum satis</u> (to levels consistent with Good Manufacturing Practice) in a number of listed dessert and confectionery products that are energy-reduced or contain no added sugar, food supplements and products for particular nutritional uses, and in sauces and mustard.

Intense sweeteners:	E 950	Acesulfam-K
	E 951	Aspartame
	E 952	Cyclamic acid and its sodium and calcium salts
	E 954	Saccharin and its sodium, potassium and calcium salts
	E 957	Thaumatin
	E 959	Neohesperidine DC

The intense sweeteners are categorised separately, with individual authorisations and maximum levels of use. The main food uses are indicated in Table 1.

<u>Progress of the Sweeteners Directive</u>

The adoption of EC Directives by the co-operative procedure can be summarised as follows. Directives are drawn up by the Commission and early drafts are discussed by the Foodstuffs Working Group or one of its sub-groups, with representatives from trade, consumers, enforcement authorities and other interested parties. When the main points are agreed, the draft is presented to the EC Council and is published in the C series of the Official Journal of the European Communities. The views of the

Table 1 Proposed EEC Authorisations for Intense Sweeteners in Main Classes of Foodstuffs

Foodstuffs	EEC Number and Name of Sweeteners					
	E 950 Acesulfam-K	E 951 Aspartame	E 952 Cyclamic acid and Na and Ca salts	E 954 Saccharin and its Na, K, Ca salts	E 957 Thaumatin	E 959 Neohesperidine DC
Water- and fruit juice-based flavour drinks[A,B] mg/l	350	600	400	80	-	30
Milk- and milk-derivative-based drinks[A,B], mg/l	350	600	400	80	-	50
Water-, milk-, fruit and vegetable-, egg-, cereal-, fat-based-desserts[A,B], mg/kg	350	1,000	250	100	-	50
Savoury snacks, mg/kg	350	500	-	100	-	-
Edible ices[A,B], mg/kg	800	800	250	100	-	50
Canned or bottled fruit[A,B], mg/kg	350	1,000	1,000	200	-	50
Jams, jellies and marmalades[A], mg/kg	1,000	1,000	1,000	200	-	50

Foodstuffs	E 950 Acesulfam-K	E 951 Aspartame	E 952 Cyclamic acid and Na and Ca salts	E 954 Saccharin and its Na, K, Ca salts	E 957 Thaumatin	E 959 Neohesperidine DC
Fruit and vegetable preparations[A], mg/kg	350	1,000	250	200	-	50
Sweet-sour preserves of fruit and vegetables, mg/kg	200	300	-	160	-	100
Confectionery[B], mg/kg	500	1,000	500	500	50	100
Starch-based confectionery[A,B], mg/kg	1,000	2,000	500	300	-	150
Sandwich spreads[A,B], mg/kg	1,000	1,000	500	200	-	50
Chewing gum[B], mg/kg	2,000	5,500	1,500	1,200	50	400
Dietetic fine bakery products, mg/kg	1,000	1,700	1,600	170	-	150
Sweet-sour preserves, semi-preserves and marinades of fish, mg/kg	200	300	-	160	-	30
Cider and perry, mg/l	350	600	-	80	-	20
Alcohol-free beer or beer with maximum 1.2% alcohol content, mg/l	350	600	-	80	-	10

Foodstuffs	E 950 Acesulfam-K	E 951 Aspartame	E 952 Cyclamic acid and Na and Ca salts	E 954 Saccharin and its Na, K, Ca salts	E 957 Thaumatin	E 959 Neohesperidine DC
Other beer, mg/l	350	600	-	80	-	10
Sauces, mg/kg	350	350	-	160	-	50
Mustard, mg/kg	350	350	-	320	-	50
Complete meal replacement formulae for weight control, mg/kg	450	800	400	240	-	100
Complete formulae and nutritional supplements, mg/kg	450	1,000	400	200	-	-
Vitamin and dietary preparations, mg/kg	2,000	5,500	-	1,200	400	-

A. Energy-reduced
B. With no added sugar

European Parliament and the Economic and Social Committee are sought at this stage. If agreement is reached a 'Common Position' is adopted by the Council by qualified majority voting. Then the European Parliament is required to give a second opinion when the proposal may be approved or rejected or proposed amendments suggested. If approved and agreed, the Council will adopt the directive by qualified majority voting; if agreement is not reached unanimous voting is required to override the opinion of the European Parliament.

Discussions are often long drawn out at the various stages. In fact, the sweeteners proposal proceeded fairly quickly to Common Position status. However, in May 1992, the European Parliament rejected the Common Position. This centered on a footnote that had been inserted after the first reading of the European Parliament, which would have allowed Member States to prohibit the use of sweeteners in non-alcoholic and low-alcohol beers brewed on their own territory following a traditional process, provided such a provision existed at the time of adoption of the directive. Parliament argued that such a restriction was against the spirit of the additives framework directive, which was intended to lead the way towards harmonised legislation between Member States. Since unanimous voting was now required before the draft could proceed and this was not reached by Council, the Draft fell; likewise, the proposal on other food additives (preservatives, antioxidants and others), which included the same footnote, also fell; both would be required to start the whole process again.

A new proposal for sweeteners was produced by July 1992[6]. The new draft contained some small detailed amendments, but the issue of use of additives in 'traditional' foods has been addressed by a proposed amendment to the text of the framework additives Directive[7]. This would apply to all three specific directives and would cover all foodstuffs produced according to a traditional process and not just beer. Discussions have since centered on an acceptable definition of 'traditional', a procedure by which a list of traditional foods would be drawn up and protection of freedom to manufacture and sell similar non-traditional foodstuffs that comply with the additives directives in the Member State/s that operate the derogation. Currently the Belgian Presidency is trying to obtain agreement both on the amendment to the framework directive and on the sweeteners proposal, even if these are approved to the Common Position in advance of the other proposals. The outcome of the later 1993 discussions will be eagerly awaited by manufacturers, and it is hoped that full agreement and an end to the discussions on the proposal are in sight.

Apart from the adoption procedure, it will be observed that the EC proposes some sweeteners that are not currently permitted in the United Kingdom. Ministers have, however, agreed to accept the use of neohesperidine DC, which was not included in the 1983 regulations as, at that time, manufacturers expressed little interest in its use. As yet, however, Ministers are not content with safety data regarding cyclamate; COT has set a temporary ADI of 0-1.5 mg/kg bw/day, which is too low to allow the use of cyclamate. Safety data regarding saccharin and aspartame have also been reviewed and in general their continued use in foods is acceptable. It is likely that new food uses as well as new sweeteners will become recommended even when the directive is adopted.

It is also of interest that acesulfam-K, aspartame, thaumatin and neohesperidine DC are also listed in the proposal that encompasses preservatives, antioxidants and other food additives; it is proposed that they should be allowed in chewing gum and a few other listed foods, with upper limits, as flavour enhancers.

5 CONCLUSION

Developing legislation today follows changes in technology and dietary recommendations much more closely than was the case a few years ago. The incorporation of newer sweeteners such as aspartame and acesulfam into food laws on a national basis was some indication that the legislators were willing to accept new evidence of a need for these substances to assist sugar reduction, for whatever reason, in the diet. The terminology of the proposed directive - 'energy reduced or no added sugar' - indicates that, for a number of foods, the concept of 'no added sugar' is becoming more popular. While one would not predict a sugarless diet, proposed EC legislation envisages manufacture and consumption of more low-sugar and sugar-free food and drink products, anticipatively in a Community market where the goals of freedom of trade, adequate protection of the consumer and informed consumer choice in the marketplace have been achieved.

REFERENCES

1. The Artificial Sweeteners in Food Regulations 1969 (S.I. 1969 No. 1817).
2. The Miscellaneous Additives in Rood Regulations 1980 (S.I. 1980 No. 1834).
3. Food Additives and Contaminants Committee Report on the Review of Sweeteners in Food, FAC/REP/34, London, HMSO 1982.
4. The Sweeteners in Food Regulations 1983 (S.I. 1983 No. 1211, as amended by S.I. 1988 No. 2112).
5. Council Directive of 21 December 1988 on the approximation of the laws of the Member States concerning food additives authorised for use in foodstuffs intended for human consumption (89/107/

EEC), Off. J. European Communities, 1989, 32 (L40), 27.

6. Proposal for a Council Directive on sweeteners for use in foodstuffs, COM(92)255 final - SYN 423, Off. J. European Communities, 1992, 35 (C206), 3.
7. Amended proposal for a Council Directive amending Directive 89/107/EEC on the approximation of the laws of the Member States concerning food additives intended for human consumption COM(93)289 final - SYN 422, Off. J. European Communities, 1993, 36 (C191), 7.

Sugar Eating Habits of Children and Adults

A. J. Adamson

COMMUNITY DIETITIAN, ENFIELD COMMUNITY CARE TRUST,
ST ANN'S HOSPITAL, ST ANN'S ROAD, LONDON N15 3TH, UK

1 INTRODUCTION

This paper will explore some of the available data on the sugar eating habits of children and adults. This is a wide remit which I propose to break into four main sections:

- Historical trends in sugar availability and consumption nationally.
- Sugar intake at different stages of life, that is, sugar intake by different age-specific groups.
- The food sources of sugars in the diets of adults and children. Where, in food terms, does sugar appear in our diets ?
- Sugar consumption at home and outside the home. With changes away from traditional eating patterns, food consumption outside the home is increasing. What effect does this have on sugar intake ?

Each of the above will be covered briefly in this paper. This overview should help to set the scene for later papers which will be discussing diet and dental problems, dietary advice and strategies to reduce sugar intake.

2 NATIONAL TRENDS IN SUGAR CONSUMPTION

Two main sources of information are available for national trends in sugar intake. Data on the supplies of sugar in Britain since 1850 (Figure 1) are available from National Food Supply statistics.[1]

Sugar supplies rose steadily from a mean daily intake of less than 40 grams per person, in the 1850's,

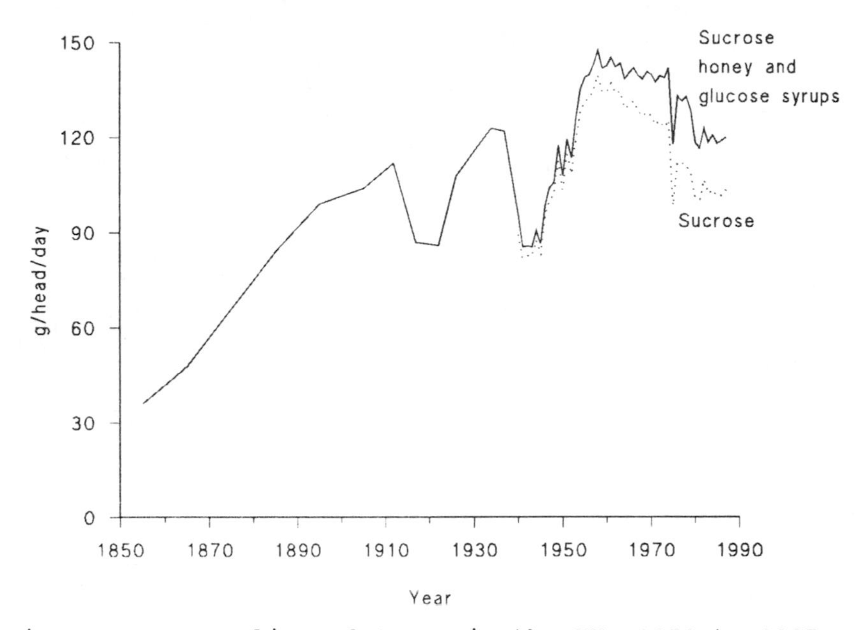

Figure 1 Supplies of Sugar in the UK: 1850 to 1987

to a peak of 148 grams per person, in 1958. The notable exceptions to this trend were during World War I and World War II when supplies fell. Supplies remained more or less constant during the 1960's and only began to fall slightly during the 1970's, to present levels of approximately 120 grams per person, per day. This accounts for approximately 16 per cent of the energy of the total food supply. Sugar supplies have changed little since 1980, present-day supplies being similar to those of 1950. Supply data makes no allowance for waste and so is likely to be an over-estimate of consumption.

The second main source of data is the National Food Survey[2,3]. This is a nationwide continuous survey of the food purchases of 7000 households throughout Britain which has been conducted by the Ministry of Agriculture, Fisheries and Food (MAFF) since 1940. The information collected includes purchases of packet sugar. Until the 1970's purchases of packet sugar (Figure 2) followed a similar pattern to that of sugar supplies - increasing post-war to peak in late 1950's early 1960's. However, the fall in packet sugar purchases continued during the 1980's and continues, while sugar supplies have remained more or less constant.

A look at the purchases of packet sugar would indicate that sugar intakes are falling, but this assumption is disputed by the sugar supply figures, which suggest that it is only the food source of the

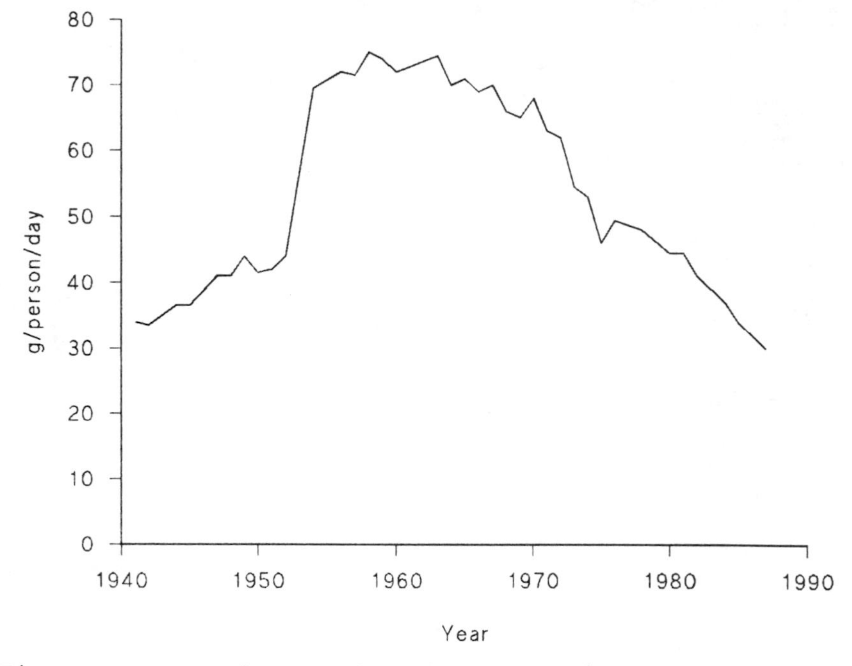

Figure 2 Purchases of Packet Sugar in the UK: 1941 to 1987

sugar that has changed. The move away from home-baking to ready-made biscuits, cakes and confectionery would help to explain this.

The National Food Survey monitors trends in food purchasing by households in Britain but does not include confectionery or food 'purchased' outside the main household food supplies (this information will be included in the National Food Survey from 1992).[3] Since 1985, data have been collected on the number of 'meals' eaten outside the home (although this figure excludes casual non-meal intakes and snacks) and on household soft drink purchasing. These data give us some clues to the changing food habits in Britain and the possible sources of sugar in that evolving diet. Between 1985 and 1990, the average number of 'meals' purchased away from the main household food supply each week increased from 3.2 to 3.8, suggesting an increase in food eaten outside the home. Figure 3 shows household soft drink purchase as measured by the National Food Survey 1985 to 1991.[4-10]

During this six-year period, purchases of soft drinks increased from approximately 800 ml to 1130 ml per person, per week. This accounts for some of the sugar supplies. It is also interesting to note the growth in the market of sugar-free 'diet' soft drinks

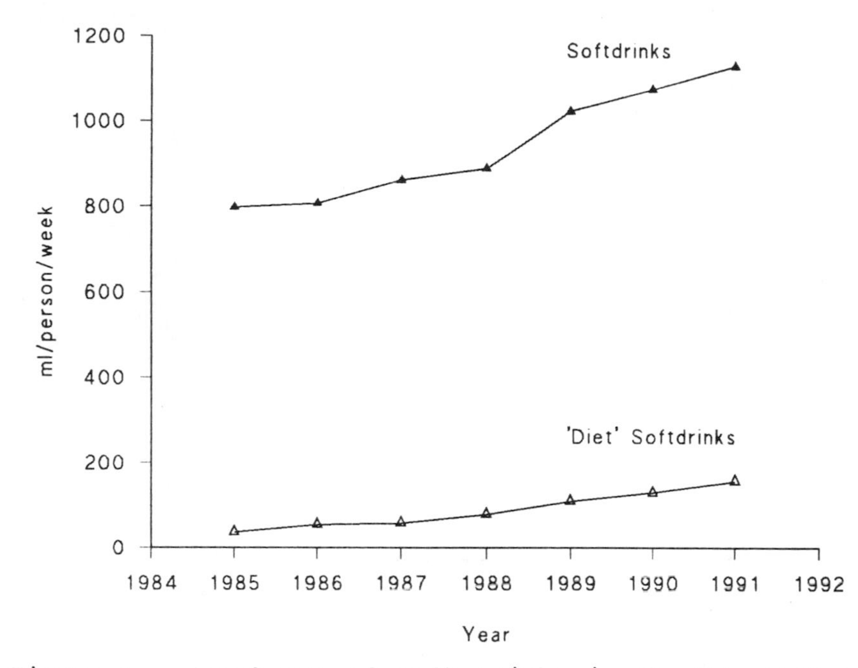

Figure 3 Purchases of Soft Drinks in the UK: 1985 to 1991. [4-10]

from 36 ml to 157 ml per person, per week (Figure 3). This indicates the potential success of an acceptable sugar-free, palatable product.

The trends in purchasing of sugar-rich foods can be compared with those of other carbohydrate-rich foods. Figure 4 shows the changes in bread purchases (of all types) between 1940 and 1990.[9] There is a continuing downward trend in bread consumption from a peak of a mean daily intake of 268 grams per person in 1948, to 114 grams per person in 1990.[9] In food terms, this represents a decrease from a mean daily intake of approximately 4 slices of bread to just 1.5 slices.

Purchasing trends for other starchy carbohydrate foods, such as potatoes, follow a similar pattern. The challenge for health education is to begin to reverse this trend.

3 SUGARS INTAKE

The National Food Survey collects data on food purchases rather than food consumption, and investigates the population as a whole rather than age-specific groups. To study sugar intake we have to look to the results of food consumption surveys.

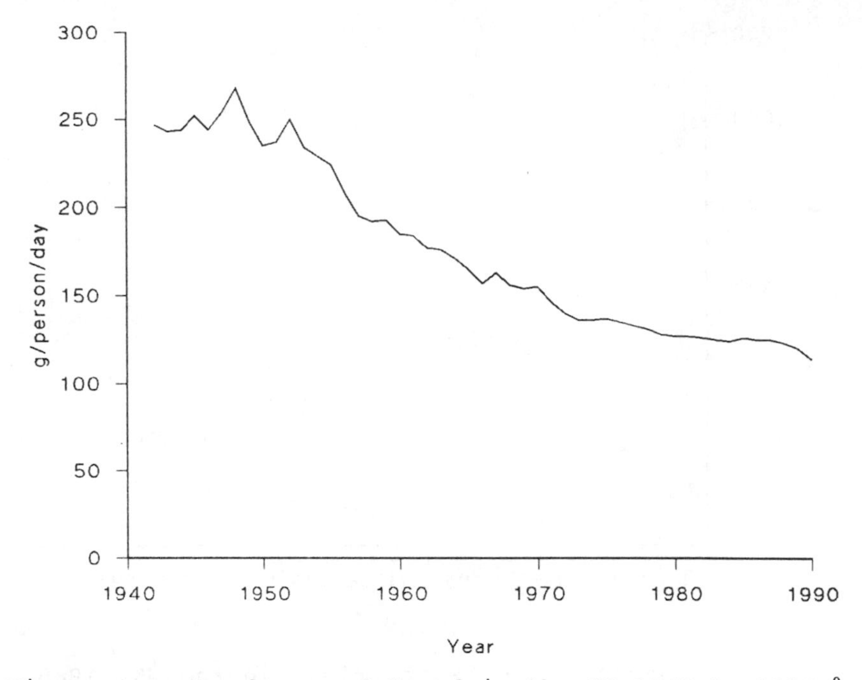

Figure 4 Purchases of Bread in the UK 1940 to 1990.[9]

Several problems exist when comparing results from different surveys, not least of which is the varying methods used to collect dietary information. Recent surveys have shown that individuals routinely under-report food intake when assessed by a 7-day weighed inventory.[11,12] Other surveys have collected data by dietary diaries.[13,14] Despite the difference in methods and the possible under-reporting of intakes, in the absence of alternative methods to measure food intake, the results of survey data provide the best information we have.

Another barrier to survey comparison is the omission of dietary sugars from analyses, and or the use of different classifications of sugars. Some surveys have measured only total carbohydrate and not sugars, others measure total sugars, whilst some have investigated 'added' and 'natural' sugars. More recently, since the 1989 COMA report on sugars,[1] the classification of sugars into non-milk extrinsic sugars, milk sugars and intrinsic sugars has been available (Figure 5).

This classification of sugar replaces the often used 'added' or 'natural' sugar and is based on the physical location of the sugar within a food or food product. This difference in physical location influences the cariogenicity of a sugar and the

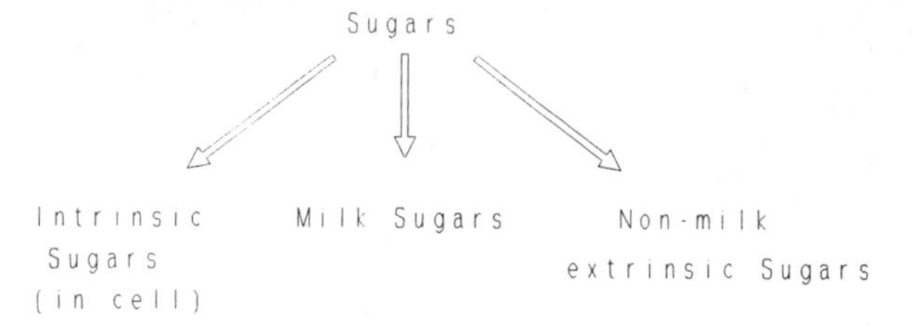

Figure 5   Classification of sugars

readiness of absorption. Milk sugar, although extrinsic is considered separately since milk sugars are less cariogenic than non-milk extrinsic sugars. Although further work is needed to ensure the consistent application of this classification, potentially its use will lead to a more consistent definition of sugars in future surveys and therefore facilitate direct comparisons.

A final problem, in the comparison of results from different surveys, is the temporal change in sugar supplies discussed above. It cannot be valid to compare directly sugar intakes reported from surveys carried out during the 1960's and 1970's with those of more recent years when sugar supplies have changed.

Despite these drawbacks, I have used the available data, from 1980 onwards, to attempt to draw a comparison of sugars intake, as a per cent of total energy intake, for different age-groups. In 1989 COMA[1] summarised the surveys investigating sugars to that date. I have used these data, with recent additions, to attempt to represent sugar eating habits through four life-stages - preschool children, schoolchildren, adults and elderly.

The first impressions are of how few data are available on the sugar eating habits of age-specific groups, particularly of the young and the elderly, and also the lack of consistent classification of sugars to allow comparison. Total and non-milk extrinsic sugar intakes, as a percentage of energy intake, are shown in Figures 6 and 7 respectively.

For pre-school children only one survey has reported sugar intake, this was a longitudinal survey of approximately 35 subjects, conducted in Cambridge between 1978 and 1982.[1] This showed that sugar contributed between 26 and 30 per cent of pre-school children's total energy intake (Figure 6).

For schoolchildren 3 surveys have reported total sugars intake at between 19 and 22 per cent of energy.[1,14] For adults more information is available, the high sugar intake of 28 per cent of energy was recorded for

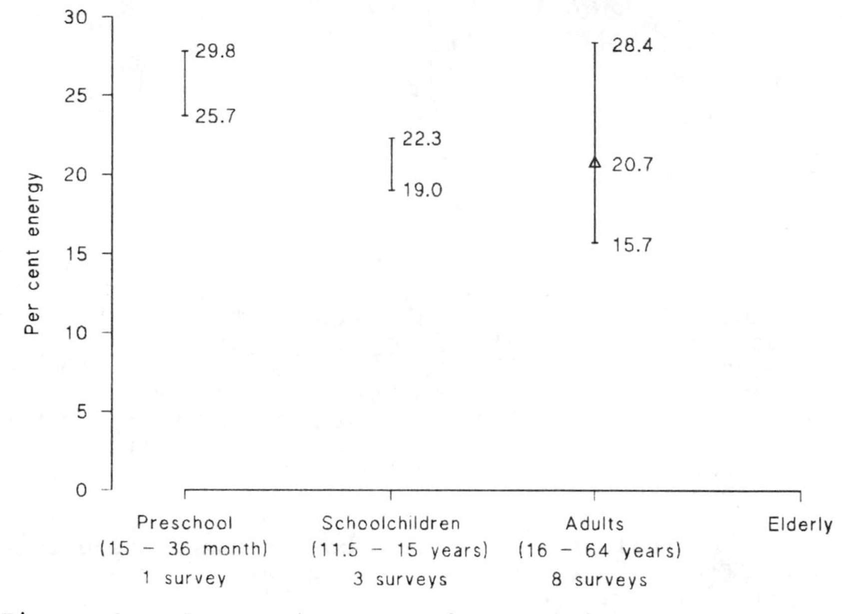

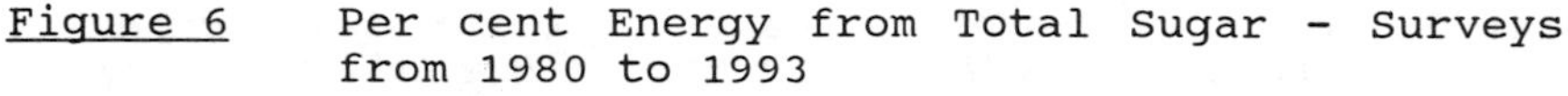

Figure 6 Per cent Energy from Total Sugar - Surveys from 1980 to 1993

a group of unemployed men in Dublin[1], whilst other surveys of adults have measured sugar intake at between 16 and 21 per cent of energy intake (Figure 6).[1,15]

Since 1980, no surveys of the elderly have investigated sugar intake, the only available data for the elderly are for 14 subjects (surveyed in 1977)[16]: this showed total sugar intakes to contribute between 16 and 25 per cent of total energy intake.

From this limited picture it is difficult to draw any conclusions about total sugar eating habits. An investigation of reported non-milk extrinsic (NME) suga intakes yields more useful information (Figure 7), although here, the number of surveys is fewer than for total sugar. Figure 7 shows the non-milk extrinsic sugars intake for different age-groups.[1,17] In 1991, the Department of Health[18] set a Dietary Reference Value (DRV) for non-milk extrinsic sugar, the recommendation being that no more than 11 per cent of food energy should be derived from this source. The DRV will be discussed in more detail in another paper. The non-milk extrinsic sugars intake measured in dietary surveys can now be compared with the DRV for non-milk extrinsic sugar (Figure 7).

For adults, only one survey since 1980 has recorded non-milk extrinsic sugar intake for both men and women - this was a survey of a group dietitians and

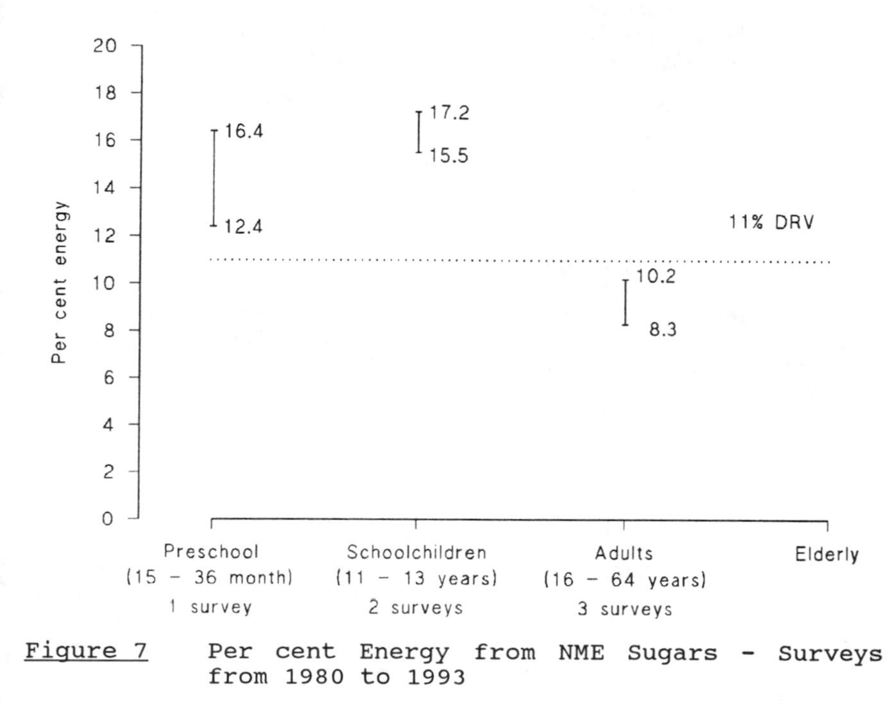

Figure 7 Per cent Energy from NME Sugars - Surveys from 1980 to 1993

their families. Two other surveys of adults were of dietitians only and pregnant women. These highly selected groups had non-milk extrinsic sugar intake within the DRV - all others exceeded 11 per cent (Figure 7). No data were available on the non-milk extrinsic sugars intake of the elderly. These data indicate that pre-schoolchildren and schoolchildren have larger intakes of non-milk extrinsic sugar than adults and that considerable changes in sugars intake are necessary before average intakes will meet the DRV set for the UK in 1991.[18]

The Department of Health is currently undertaking a survey of pre-schoolchildren in Britain and has plans to begin a survey of the food intake of the elderly. Hopefully, these surveys will include a study of total and non-milk extrinsic sugar intake and provide much needed data.

There is evidence to suggest food habits track through the generations.[3,19] Therefore eating habits may have more to do with diet in childhood than with age, this is termed the 'vintage effect'. The implication of the 'vintage effect' is that food habits formed early in life follow into adulthood. This is important for dietary education and suggests that targeting the young now, with dietary education for change is paramount if future generations are to meet dietary recommendations. Advice needs to be appropriate and relevant to the target group and therefore information on the food habits and diet of these target groups is essential.

Information on food sources of sugar is useful for the formation of positive messages in food terms, encouraging an increase in 'starchy carbohydrate' foods rather than an increase in fat or sugar.

4 FOOD SOURCES OF SUGAR

A small number of surveys have investigated either food sources of total sugar or non-milk extrinsic sugar; however a complication lies in the food groupings used - no consistent food groups exist which can lead to some differences between surveys. A step forward in this area would be agreement and consistent application of food groupings.

A survey of 2700 British schoolchildren conducted in 1983[20] originally did not investigate sugar intake; this has recently been reanalysed to look at the food sources of total sugar (but, unfortunately, not non-milk extrinsic sugar).[21] In 1987, a survey of 2000 adults in Britain[15] investigated food sources of total sugar. Using these data we can compare the role of the different food groups in the diets of the adults (Table 1) and children (Table 2).

Although food groupings differ slightly for the two surveys it can be seen that confectionery played a larger part in the children's diets (18 per cent of total sugar) than in the diets of adults (6 per cent of total sugar). However, if the foods often targeted for change are considered together - table sugar, confectionery, cakes and biscuits, soft drinks and puddings - these foods contributed similar proportions of the total sugar intake to the diets of both adults and children (63 per cent for adults and 65 per cent for schoolchildren).

Table 1 The Mean Daily Intake and Percentage Contribution of Selected Food Groups to Total Sugar Intake of Adults in 1987.[15]

Total sugar (g)	men = 115 women = 86
Food group	**%**
Sugar, confectionery, preserves	29
(of which confectionery)	6
Biscuits and cakes	12
Puddings and ice-cream	5
Beverages	17
Milk and milk products	13

Table 2 The Mean Daily Intake and Percentage Contribution of Selected Food Groups to Total Sugar Intake of Children in 1983[21].

Total sugar (g)	123
Food group	**%**
Confectionery	18
Table sugar	16
Biscuits and cakes	13
Puddings and ice-cream	9
Soft drinks	9
Milk and milk products	10

More recently, in 1990, a survey of 379 English children (aged 11 to 12 years) investigated sugars intake, including the dietary sources of both total and non-milk extrinsic sugars. The same food groupings were used as in a similar survey conducted in the same schools ten years earlier.[17] The results from these surveys show the changes in sugar eating habits from 1980 to 1990. The mean total sugar intake for these children was unchanged at approximately 117 grams a day (Table 3). Non-milk extrinsic sugar increased from 83 grams/day to 90 grams/day, whilst milk and intrinsic sugars fell from 34 to 28 grams/day. Therefore, in 1990 a larger proportion of the total sugar intake by these children was non-milk extrinsic (Table 3). In 1990 non-milk extrinsic sugar accounted for 17.2 per cent of the total energy intake of these children, an increase from 15.5 per cent in 1980; this can be compared with the DRV value of 11 per cent.

Between 1980 and 1990, changes in the food sources of total sugar were apparent (Table 4). There was a dramatic fall in intake of table sugar from 16 to 7 per cent, but an increase in intakes from both confectionery (22 to 26 per cent) and soft drinks (13 to 20 per cent) - these changes reflect national trends. For non-milk extrinsic sugar intake (Table 5) there was an increased role of confectionery (29 to 33 per cent) and soft drinks (19 to 27 per cent), whilst table sugar fell from 23 to 12 per cent. In 1990, 60 per cent of non-milk extrinsic sugar came from confectionery and soft drinks together. One other notable change in the source of non-milk extrinsic sugar was from breakfast cereals, an increase from 2 to 5 per cent of non-milk extrinsic sugar intake.

Table 3 Mean Daily Sugars Consumption of 11 to 12-year old English Children in 1980 and 1990

	1980	1990
Total (g)	117.1	118.3
Per cent energy	21.9	22.3
NME (g)	83.0	90.3
Per cent energy	15.5	17.2
Milk and Intrinsic (g)	34.1	27.9
Per cent energy	6.4	5.3

Table 4 Percentage Contribution of Selected Food Groups to Total Sugar Intake of 11 to 12-year old English Children in 1980 and 1990.

%	1980	1990	%
22	Confectionery	Confectionery	26
16	Table sugar	Soft drinks	20
13	Soft drinks	Milk	10
11	Biscuits & cakes	Biscuits & cakes	9
11	Sweet puddings	Sweet puddings	7
9	Milk	Table sugar	7

Table 5 Percentage Contribution of Selected Food Groups to Non-milk Extrinsic Sugar Intake of 11 to 12-year old English Children in 1980 and 1990.

%	1980	1990	%
29	Confectionery	Confectionery	33
23	Table sugar	Soft drinks	27
19	Soft drinks	Table sugar	12
11	Biscuits & cakes	Biscuits & cakes	11
9	Sweet puddings	Sweet puddings	6
3	Syrups & preserves	Breakfast cereals	5
2	Breakfast cereals	Syrups & preserves	2

Sugar intake should not be viewed in isolation, but rather the diet as a whole considered. Confectionery and soft drinks contributed 60 per cent of non-milk extrinsic sugar intake and so are targets for change. However, these food groups also contributed approximately 15 per cent of the total energy intake of these children and 11 per cent of their fat intake came from confectionery.[22] A reduction in the intake of confectionery could result in a favourable fall in fat intake (these children obtained 40 per cent of their energy from fat compared with the DRV of 35 per cent) but, since most of the children were not overweight, the energy intake from these food sources would have to be replaced. Table 6 shows the top six food sources of energy, protein, fat, unavailable carbohydrate and non-milk extrinsic sugars for these 11 to 12-year old children.

Table 6 The Top Six Sources of Energy, Fat, Protein, Unavailable Carbohydrate and Non-milk Extrinsic Sugars for 11 to 12-year-old English Children in 1990.[19,22]

Energy	NME sugar	Fat
Meats	Confectionery	Meats
Confectionery	Soft drinks	Confectionery
Chips	Table sugar	Spreading fats
Bread	Biscuits & cakes	Chips
Biscuits & cakes	Sweet puddings	Biscuits & cakes
Milk	Breakfast cereals	Milk

Protein	Unavailable carbohydrate
Meats	Chips
Milk	White bread
Bread	Vegetables (not pots)
Potatoes	Breakfast cereal
Fish	Crisps
Confectionery	Wholemeal bread

National data leaves us in no doubt that purchases of packet sugar are falling but sugar available for consumption remains more or less constant. The results presented above show that whilst consumption of visible forms of sugar are falling, the use of food products with 'hidden' sugar are increasing to compensate.

5 PLACE OF PURCHASE

Another piece of the picture of sugar eating habits is the question of where sugar containing foods are purchased. There is evidence that eating outside the home is increasing. The National Food Survey has estimated that expenditure on food outside the home has increased from 20 per cent of total food expenditure in 1980 to 30 per cent in 1990.[3]

Loughridge et al.[23] investigated the amount of food purchased outside the home (that is, food both purchased and consumed outside the home and therefore not part of the household food supply) by adults (16-64 years). It was found that this contributed 25 per cent of the total energy intake. Food from outside the home contained more sugar and less protein, 'fibre', iron and vitamins per unit energy than the diet as a whole (Table 7).

In 1990, food eaten 'outside the home' by 11 to 12-year old English children was investigated.[24] It was found that these children obtained 31 per cent of their energy intake from sources outside the home (Table 7). This accounted for 38 per cent of non-milk extrinsic sugars intake but less than 30 per cent of calcium and iron intake (Table 8).

Table 7 Per Cent of Selected Nutrient Intake Obtained from Sources 'Outside the Home' by Adults[23] and Children.[24]

	Adults (16-64 years)	Children (11-12 years)
Energy	25	31
Protein	22	25
Fat	25	31
Carbohydrate	24	32
Sugar	26	34
'Unavailable carbohydrate'	17	25

The food purchased by these children was further analysed, food bought at school meals and from a shop or cafe was measured. The shop/cafe place of purchase was particularly interesting with respect to sugar intake, these children obtained 12 per cent of their total energy intake from this source but 22 per cent of their non-milk extrinsic sugar intake and only 8 per cent of calcium, iron (Table 8).

Table 8 The Per Cent of Selected Nutrient Intake Obtained from Sources 'Outside the Home' by 11 to 12-year old English Children[24]

	All sources 'outside the home'	Shop/cafe only
Energy	31	12
Total sugar	34	18
NME sugar	38	22
Fat	31	11
Calcium	27	8
Iron	28	8

Food purchased by children is outside parental control and therefore, advice for change, needs to be directed at the children themselves. The accessibility (in the tuck-shops frequented by children) of acceptable alternatives, at an affordable cost, is an area for improvement. Where reduced-sugar alternatives do exist these are often not marketed at the young and not stocked by shops local to schools.

6 CONCLUSIONS

Despite health education messages, total sugar intake has remained unchanged for over a decade. Whilst intakes of 'visible' sugar intake are falling, intakes of confectionery and soft drinks have increased to compensate. Results from a survey of 11 to 12-year olds in 1980 and 1990, show that non-milk extrinsic sugar intake has increased and, at 17 per cent of energy intake in 1990, is above the recommended maximum of 11 per cent.[18]

Confectionery and soft drinks are major contributors to non-milk extrinsic sugar intake and should be the targets for reduction. However, there is a need for positive messages and acceptable alternatives, rather than simply messages to reduce sugar intake. Later papers will be discussing dietary advice and sugar replacement.

ACKNOWLEDGEMENTS

Information on the diets of 11 to 12-year-old Northumbrian children was collected through grants from the Medical Research Council and Newcastle Health Authority. This manuscript was prepared with the assistance of P.M. Adamson.

REFERENCES

1. Department of Health, 'Dietary Sugars and Human Disease.', HMSO. London, 1989.
2. Ministry of Agriculture, Fisheries and Food, 'Household food consumption and expenditure: Annual report of the National Food Survey Committee.', HMSO. London, 1947-1987.
3. Ministry of Agriculture, Fisheries and Food, 'Fifty years of the National Food Survey 1940-1990.', HMSO. London, 1991.
4. Ministry of Agriculture, Fisheries and Food, 'Household food consumption and expenditure 1985: Annual Reports of the National Food Survey Committee.', HMSO. London, 1987.
5. Ministry of Agriculture, Fisheries and Food, 'Household food consumption and expenditure 1986: Annual Reports of the National Food Survey Committee.', HMSO. London, 1987.
6. Ministry of Agriculture, Fisheries and Food, 'Household food consumption and expenditure 1987: Annual Reports of the National Food Survey Committee.', HMSO. London, 1988.
7. Ministry of Agriculture, Fisheries and Food, 'Household food consumption and expenditure 1988: Annual Reports of the National Food Survey Committee.', HMSO. London, 1989.
8. Ministry of Agriculture, Fisheries and Food, 'Household food consumption and expenditure 1989: Annual Reports of the National Food Survey Committee.', HMSO. London, 1990.
9. Ministry of Agriculture, Fisheries and Food, 'Household food consumption and expenditure 1990 with a study of trends over the period 1940-1990: Annual Reports of the National Food Survey Committee.', HMSO. London, 1991.

10. Ministry of Agriculture, Fisheries and Food, 'Household food consumption and expenditure 1991: Annual Reports of the National Food Survey Committee.', HMSO. London, 1992.
11. M.B.E. Livingstone, A.M. Prentice, J.J. Strain, W.A. Coward, A.E. Black, M.E. Barker, P.G. McKenna and R.G. Whitehead, Br. Med. J., 1990, 300, 708.
12. M.B.E. Livingstone, P.S.W. Davies, A.M. Prentice, W.A. Coward, A.E. Black, J.J. Strain and P.G. McKenna, Proc. Nut. Soc., 1991, 50, 15A.13.
A.F. Hackett, A.J. Rugg-Gunn, M. Allinson, D.R. Appleton and J.E. Eastoe, Br. J. Nutr., 1984, 51, 347.
14. A. Adamson, A. Rugg-Gunn, T. Butler, D. Appleton and A. Hackett, Br. J. Nutr., 1992, 68, 543.
15. J. Gregory, K. Foster, H. Tyler and M. Wiseman, 'The dietary and nutritional survey of British adults.', HMSO. London, 1990.
16. S. Bingham, N.I. McNeil and J.H. Cummings, Br. J. Nutr., 1981, 45, 23.
17. A.J. Rugg-Gunn, A.J. Adamson, D.R. Appleton, T.J. Butler and A.F. Hackett, J. Hum. Nutr. Diet., 1993, 6, 419.
18. Department of Health, 'Dietary Reference Values for Food Energy and Nutrients for the United Kingdom.', HMSO. London, 1991.
19. A. Must, P.F. Jacques, G. Dallal, C.J. Bajema and W.H. Dietz, New Engl. J. Med., 1992, 327, 1350.
20. Department of Health, 'The Diets of British Schoolchildren.', HMSO. London, 1989.
21. S.A. Gibson, J. Hum. Nutr. Diet., 1993, 6, 355.
22. A.J. Adamson, A.J. Rugg-Gunn, D.R. Appleton, T.J. Butler and A.F. Hackett, J. Hum. Nutr. Diet., 1992., 5, 371.
23. J.M. Loughridge, A.D. Walker, H. Sarsby and R. Shepherd, J. Hum. Nutr. Diet., 1989, 2, 361.
24. A.J. Adamson, A.J. Rugg-Gunn, D.R. Appleton, T.J. Butler, 1993, Unpublished.

Problem Areas in Liquid Oral Medication

A. Maguire

DEPARTMENTS OF CHILD DENTAL HEALTH AND ORAL BIOLOGY,
THE DENTAL SCHOOL, UNIVERSITY OF NEWCASTLE UPON TYNE,
FRAMLINGTON PLACE, NEWCASTLE UPON TYNE NE2 4BW, UK

1 INTRODUCTION

Liquid oral medicines have been around for a long time. The oral route for the administration of drugs has the longest history of use and liquids were routinely used well before the advent of tablets in 1843 when William Thomas Brockedon was granted a patent for a tabletting machine. However until recently sugars have almost exclusively been used to sweeten liquid medicines to make them more palatable and so aid compliance, primarily in paediatric and geriatric medicine.

The aetiology of dental caries is well researched and documented[1,2,3] and the primary relationship of host, micro-organisms, substrate and time forms the basis for the current concept of caries aetiology, now thoroughly accepted in scientific circles. The effect of diet and, in particular, fermentable carbohydrates, in providing a substrate from which micro-organisms produce acid is well researched[4,5] and therefore the use of sugar in medicines, especially those taken frequently and long-term is of obvious concern to all health professionals.

The COMA Report

The COMA Report[6], published in 1989, looked at dietary sugars and human disease, and considered the impact of sugars in medicines in relation to dental caries. The terms of reference of the panel were " To examine the role of dietary sugars in human disease and make recommendations". Of these, number 8 stated that "An increasing number of liquid medicines are available in sugar-free formulations. When medicines are needed, particularly long-term, sugar-free formulations should be selected by parents and medical practitioners. The Panel *recommends* that government should seek the means to reduce the use of sugared liquid medicines." In view of this it is important to look at the function of sugar in medicines, and try to identify problem areas which may be encountered in seeking to implement this COMA report recommendation.

2 WHAT ROLE DOES SUGAR HAVE IN THE FORMULATION OF LIQUID MEDICINES?

Sugars have a number of properties which are exploited in the formulation of liquid medicines (Table 1).

Table 1 Uses of sugars in liquid medicines formulation

DILUENT
PRESERVATIVE
SWEETENING
WETTING
DEMULCENT
VISCOSITY MODIFYING

As a diluent

Syrup BP, containing 667g of sucrose made up to 1000g with purified water, has, until very recently, frequently been used as a diluent when preparing a liquid medicine for the use of a younger patient or for someone not requiring the full dosage supplied by the manufacturer. However, in 1992 in the U.K., "because of the problems of giving relatively large volumes to neonates, concern about the stability of diluted medicines and the incidence of dental caries in children", the Department of Health decided that the convention of diluting medicines should end, and since July 1992 community pharmacists have no longer been expected to dilute fractional doses of medicines for small children to allow administration from a standard 5 ml spoon[7]. Instead they have been dispensing the medicine undiluted with an oral syringe and a rubber adaptor (Figure 1) which has to be fitted to the medicine bottle to allow the dose to be withdrawn. Once the dose is retrieved, it is adjusted and checked and then given slowly to the child to prevent gagging or choking with the tip of the syringe just inside the child's mouth pointing towards the inside of the cheek.

This ending of the dilution convention is welcome as far as dental health is concerned; however the proportion of liquid oral medicines that this applies to is really quite small (Table 2). Of all the liquid oral medicines dispensed on prescription during a one year period in 1987, these diluted medicines accounted for 280.3 thousand prescriptions, and 37.8 thousand litres out of a total of 44 million prescriptions and 10 million litres of liquid oral medicines dispensed in

Great Britain. Therefore the impact of ending the dilution convention will be small, but it is a welcome start.

As a preservative

Preservatives are often necessary to maintain the physical attributes of a liquid pharmaceutical product, and thus avoid any hazard to health. An ideal preservative is active against a wide range of micro-organisms, is physically, chemically and biologically stable for the lifetime of the product, and non-toxic, non-sensitising, adequately soluble, taste compatible and odour acceptable in relation to the product[8].

As a 66.7% w/w solution, sucrose has preservative properties, although most syrups and liquids have additional preservatives such as parahydroxybenzoic acid esters and salts of benzoic and sorbic acid included to prolong their shelf life.

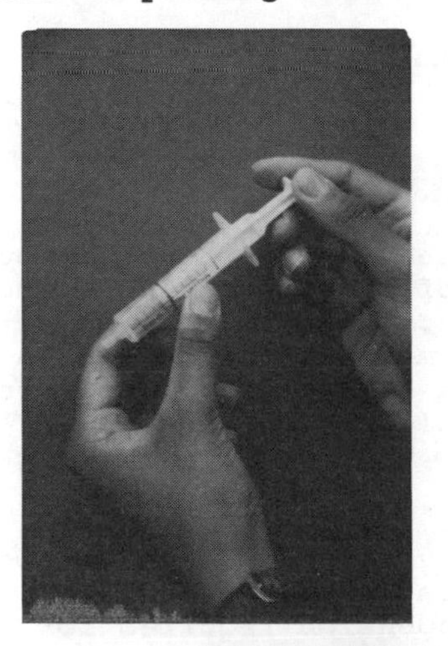

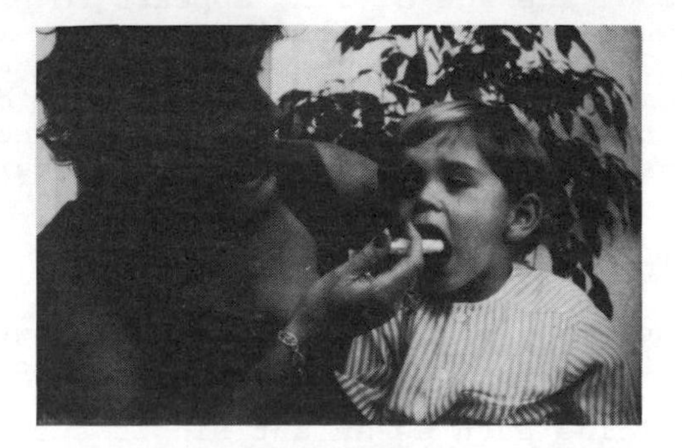

Figure 1. Use of the oral syringe

Table 2. Prescription analysis. Number of prescriptions (1000s) and quantity (litres) of (i) all liquid oral medicines (LOMs), (ii) long-term LOMs, and (iii) LOMs diluted with syrup, dispensed in Great Britain during a one year period (1987).

	Prescriptions (1000s)	Quantity (Litres)	Number of preparations
All LOMs	43,901	10,323,000	581
LOMs suitable for children	38,020	6,589,000	453
Long-term LOMs	22,736	3,844,000	196
LOMs dispensed diluted with syrup	280	37,821	23

As a sweetening agent

The sweetening properties of sugar are obvious. However, the taste masking of a drug to make it more palatable is a complex science since the taste of a drug in liquid form is much more profound than in solid form. Therefore to provide a suitably palatable form of liquid paracetamol for example, which has a particularly foul taste, may require more masking than other drugs in liquid form.

A number of sugar-free versions of paracetamol oral suspension 120mg/5ml are available. However, the 250mg/5ml of paracetamol suspension required for children over the age of six, appears to present problems when trying to formulate a sugar-free version, particularly in achieving a product which will pour easily from the bottle. In spite of these difficulties there are now a number of sugar-free paracetamol suspensions available for the over six-year age group.

As a wetting agent

Sucrose is used in liquid form as a wetting agent to reduce the surface tension between insoluble solids which need to be suspended uniformly throughout a liquid medicine such as an antibiotic suspension, so that the patient receives the same amount of active ingredient with the first dose as the last dose.

As a demulcent and viscosity modifying agent

Sugars in liquid medicines have soothing or demulcent properties which are used in the formulation of cough suppressants and anti-tussives such as pholcodine linctus which is available as a sugar-containing linctus and as a sugar-free liquid alternative, Pavacol-D which contains the intense sweetener saccharin and the polyol sorbitol as sweetening agents. The viscous nature of the cough syrup soothes the throat as it slides down, and improves the pourability of the liquid medicine, enabling it to be poured from the bottle with more control.

Therefore, sugars have a number of properties, exploited in the formulation of liquid medicines, and the task of replacing the sugars to provide a suitable range of non sugar-based alternatives can be difficult.

3 LIQUID ORAL MEDICINE CONSUMPTION

Introduction

A series of surveys carried out here in Newcastle has examined the overall use of liquid and syrup medicines particularly in children[9,10]. This included long-term prescribed use under the care of consultant paediatricians and shorter term use through over-the-counter purchase of medicines.

These surveys also investigated the sugars content of these medicines and looked at the dental health of children taking them. Some of the results of these and other studies and some problems they have highlighted are given below.

What is the size of the problem?

It is useful to consider this in terms of the prevalence of use of liquid oral medication, and its impact upon dental health.

Prevalence. General medical and dental practitioners in Great Britain issue over 400 million scripts per year to their patients of all ages. This is quite conservative when compared with the rest of the European Community and the United States (Table 3). In the United Kingdom this amounts to an average of 7 scripts per person per year, obviously higher in the paediatric and geriatric age groups.

Some part of the disparity between prescribing rates in different countries can be explained by differences in medical opinions and social perceptions of disease. In addition, the extent of the active role the dispensing pharmacist plays in providing diagnosis and treatment over-the-counter, will influence differences in prescribing rates between countries.

Pharmaceutical spending consumes approximately 0.5% of the Gross Domestic Product in the UK annually, compared with an average outlay of 0.9% in Western Europe (Table 4). The £64 spent per person in the UK in 1990 was 23% below the Western European average of £83 per person, with spending in France and Switzerland exceeding that in the UK by at least 50%[12].

Table 3. Prescribing rates - International comparison[11]

Country	Number of scripts/patient/year
USA	16.6
Italy	11.3
West Germany	11.2
France	10.1
Spain	9.6
New Zealand	8.5
Australia	7.7
UK	7.0
Sweden	4.7

In the UK, liquid oral medicines account for approximately 11% of the 401 million prescriptions dispensed. In 1987 this represented 10 million litres of liquids prescribed and dispensed in Great Britain (Table 2). In addition to this prescribed use, there is, of course, large-scale use of liquid medicines purchased over-the-counter (OTC medicines), many of which are taken on a regular basis, for example cough medicines and analgesics.

The prevalence of long-term use of liquid oral medication in 5 districts (Newcastle, Gateshead, North Tyneside, South Tyneside, and Northumberland) in the Northern Region of England was investigated (Figure 2).

Through a survey of consultant paediatricians and general medical practices the prevalence of use was approximately 1:357 children. At any one time 630 1-16 year olds were taking a prescribed liquid oral medicine long-term, that is, daily or on alternate days for at least 3 months. Extrapolating these figures allows us to estimate that for the whole of the Northern Region with a population of 0.62 million 1-16 year olds, 1764 children will be taking long-term liquid oral medication on prescription. On a national basis, in a population of 55 million, this represents nearly 30,000 children (Figure 3).

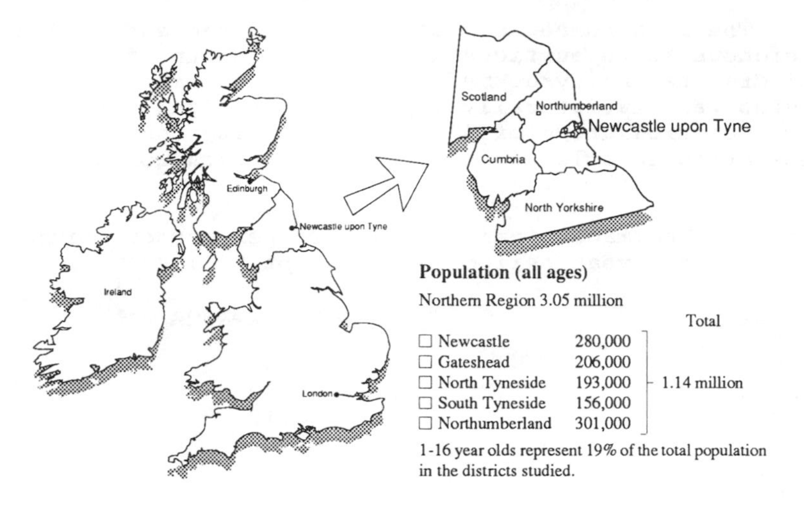

Figure 2. The 5 Districts studied in the Consultant Survey

Dental Health. From the dental point of view, it is the pattern of use of these sugared liquid oral medicines, particularly their chronic use in children, which gives cause for concern.

The effects of sugar-based liquid oral medicines on dental health are well documented; Roberts and Roberts[13] in their study of children attending hospital outpatient departments demonstrated detrimental effects upon dental health through long-term use (Table 5).

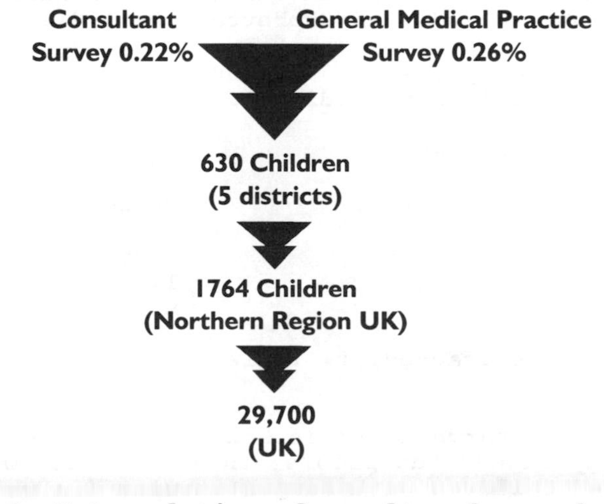

Figure 3. Extrapolation of results of prevalence surveys.

The mean number of decayed, extracted and filled deciduous tooth surfaces in the study group of 44 children taking syrup medicines daily for more than 6 months was statistically significantly higher than that for a control group of 47 children taking their medication in solid dose form or no medication at all.

Table 4. Pharmaceutical expenditure per person during a one year period (1990 except *: 1989)[12]

Country	£ per person
Japan*	198
France	112
USA*	110
Italy	107
Switzerland	106
Germany	91
Belgium	90
Finland	89
Sweden	77
Austria	71
Spain	64
UK	64
Denmark	61
Norway	56
Netherlands	53
Portugal	46
Ireland	39
Greece	32

Table 5. Dental disease in chronically sick children (Roberts G.J., and Roberts I.F. 1981)[13]

44 study	Daily LOMs > 6 months
47 control	Tablets/no medication

Age: 9 months - 6 years

	Study	Control
mean def(s)	5.55	1.26
mean df(s)	3.97	1.26
mean e(s)	0.34	nil

d= decayed, e= extracted, f= filled, s= tooth surfaces

Another research group in Canada, Kenny and Somaya[14], approached the problem in a different way. They studied children with rampant caries (in which many teeth decay rapidly and the tooth surfaces involved are often those not normally prone to decay), and looked at the medication history of 20 chronically sick children identified with this gross decay. Of the children studied 8 had epilepsy, 6 a cardiac condition, 4 infections, 1 spina bifida and the remaining 1 arthritis. Their median age was 31 months and their mean daily supplemental sugar intake from liquid oral medication was 17g with a range between 1 and 46 g. 44 different liquid drugs were used and of the 10 most commonly prescribed medications, all had a sucrose content of between 12% and 80% in each 5 ml dose. The study confirmed suspicions that oral liquid medications when used over long periods are regularly dispensed into the mouths of sleeping children, which is a time when salivary flow is at its lowest and a longer and more profound drop in plaque pH more likely, and therefore the chances of decay occuring very much higher.

The Dental Health Survey carried out here in the North-east of England[15] looked at children between the ages of 2 and 17 years who had been taking liquid oral medication of a single sweetener type (that is, sugar-based or sugar-free) for at least one year and their siblings. The results of this Survey show that differences were most marked in the deciduous dentition. In the sugar-based study group the total number of decayed, missing, and filled deciduous tooth surfaces (dmfs) for the sugar-based study group was 4.6 compared with 2.5 in the sugar-free study group (Figure 4). The siblings were quite similar with an average of 3.5 and 3.2 surfaces involved. When these figures were broken down into missing tooth surfaces (which were due in the vast majority of instances to teeth having been extracted) and decayed and filled tooth surfaces (dfs), and expressed as a % of the total dmfs, the missing surfaces accounted for 37% of the total for the sugar-based and 28% of the total for the sugar-free study groups respectively, compared with 17% and 11% of the sibling control groups.

When the differences according to tooth type were considered (Figure 5), the decayed, missing and filled incisor and canine teeth formed 31% of the total for the whole mouth for the sugar-based study group compared with 1% in their siblings. In the sugar-free study group, anterior tooth caries accounted for 15% of the total for the whole mouth and its sibling group 22%. The results were analysed by Ordinal Logistic Regression Analysis, a statisical method designed to determine which independent explanatory variables had the most influence on each dental index measured allowing for the effects of possible confounding factors such as age and exposure to fluoride.

Analysis showed the effect of group classification (that is sick children [sugar-based study and sugar-free study] versus healthy children [the two sibling groups]) was statistically significant for decay in incisor and canine teeth at the 5% level (p=0.0463).

These detrimental effects are obviously very relevant and especially when the children involved are chronically sick and in whom certain dental procedures such as dental extractions under general anaesthetic can carry greatly increased morbity.

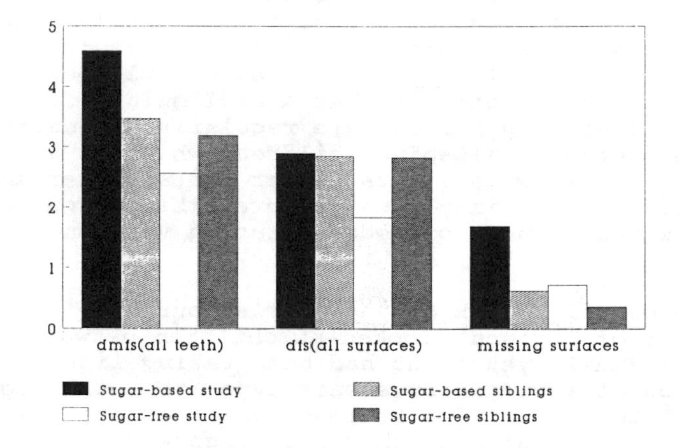

Figure 4. Dental health survey - missing deciduous teeth

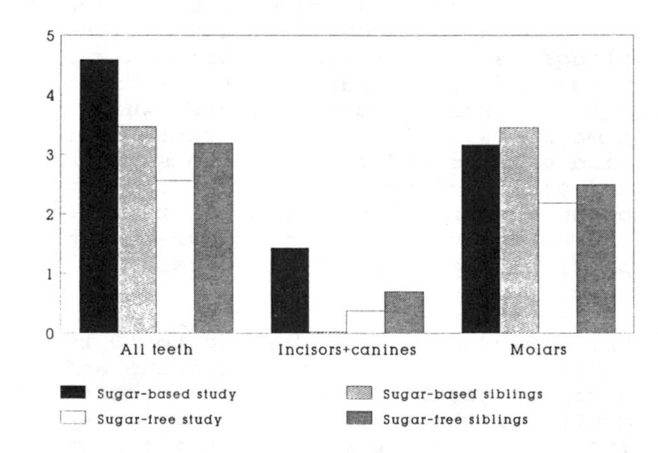

Figure 5. Dental health survey - mean dmfs by tooth type

4 WHO TAKES LIQUID ORAL MEDICATION LONG-TERM AND WHICH MEDICINES ARE INVOLVED?

The study of children under the care of consultant paediatricians identified 243 children taking liquid medicines long-term. The most frequent chronic medical problems identified were Epilepsy, Cystic Fibrosis,

Chronic Renal Failure, Constipation, Asthma and Recurrent urinary tract infections (Table 6).

Table 6. The ten most common medical problems for which long-term oral medication was prescribed.

Medical problem	Number of children	%
Epilepsy	58	24%
Cystic Fibrosis	43	18%
Chronic renal failure	27	11%
Chronic constipation	20	8%
Asthma	18	7%
Recurrent urinary tract infections	13	5%
Cardiac defects	13	5%
Vesico-ureteric reflux	6	3%
Atopy	5	2%
Enuresis	4	2%
Others	36	15%
Total	243	100%

Table 7 identifies the nature of the sweetening agents contained in liquid oral medicines taken long-term for these chronic medical conditions: the variable drugs are those in which, depending on how specifically they are prescribed , either a sugar-free or a sugar-based medicine is dispensed to the child.

The anti-convulsant agent sodium valproate is available in a sucrose-containing syrup and a sugar-free liquid containing sorbitol and saccharin. These have equivalent bioavailabilities, that is, a similar dose produces the same therapeutic response. However, they are dispensed in vastly differing amounts; nearly three times as many prescriptions for syrup as liquid being dispensed over a one year period[16]. Children with cystic fibrosis are routinely managed with liquid long-term therapy; mainly antibiotics and vitamin replacements.

5 SPECIFICITY OF PRESCRIBING

How specifically a medicine is prescribed is important if we are to move towards more sugar-free prescribing and dispensing, and here I am deliberately distinguishing between prescribing ie. what is written on the script taken to the pharmacist and what is actually dispensed by the pharmacist.

The anti-infective agent cotrimoxazole, used long-term in recurrent urinary tract and chest infections, was defined as containing a variable sweetener in Table 7. Prescribed by a proprietary name, Septrin, it contains a combination of saccharin plus sorbitol. Prescribed by the generic name, cotrimoxazole paediatric suspension BP, it contains between 44-60% sucrose in a 5ml dose depending on the generic manufacturer and there are at least 8 of these in the UK.

This aspect of specificity of prescribing was highlighted in the Consultant Paediatrician Survey (Table 8).

Of the 381 instances of liquid oral medicine prescribing in the 243 children identified as taking long-term liquid oral medication, 122 instances of prescribing were for sugar-based medicines, 143 for non-sugar-based medicines, and 116 for variable medicines. However, of those sugar-based medicines, 51% were prescribed by generic name, 49% by proprietary name, whereas, of the non-sugar-based medicines, 22% were prescribed by generic name, 78% by proprietary name.

As national drugs bills spiral ever upwards - the cost of drugs in the NHS for 1992 was over 3 billion pounds - the government in the UK has introduced a number of changes to try to control costs. Generic prescribing by doctors and dentists has been encouraged for a number of years now but increased dramtically in 1985 with the introduction of limited lists of drugs available for use in the NHS.

The graph in Figure 6 shows the changes in generic prescribing and dispensing between 1970 and 1992 and predictions for the year 2000, when generic prescribing is expected to level out at approximately 50% of all prescriptions written. Generic dispensing actually occurs at a lower level since not all drugs prescribed by their generic name are actually available generically. In addition, for stock control purposes, some pharmacists may consiously only stock a proprietary drug which can then meet prescription demands for both proprietary and generically prescribed drugs, although they may only be reimbursed for the cost of the generic product.

Other countries have to contend with the concept of Generic Substitution as a cost cutting measure, whereby prescribed proprietary name drugs can be substituted with cheaper generics by the dispensing pharmacist. This is actually occurring in some EC countries and, as such, means that the prescribing clinician has little control over the drug that the patient receives. Fortunately we have not reached that stage in the UK, with the government at the moment

preferring to encourage generic prescribing and limiting the use of certain drugs by black and white listing to control overall costs.

Table 7. Liquid oral medicines used long-term by medically compromised children

SUGAR-BASED	VARIABLE	SUGAR-FREE
Chronic Renal Failure		
Calcium carbonate	Cotrimoxazole	Sodium chloride
Sodium bicarbonate	Nalidixic acid	Potassium chloride
Ferrous sulphate		One α Vit D_3
Nutritional supplement		Nitrofurantoin
Cephalexin		Trimethoprim
Recurrent Urinary Tract Infections		
Cephalexin	Cotrimoxazole	-
	Amoxycillin	
	Nalidixic acid	
Constipation		
Lactulose	-	-
Sennosides		
Docusate sodium		
Asthma		
-	Salbutamol	Ketotifen
		Terbutaline
Malignant Disease		
-	Cotrimoxazole	
Cardiac Lesions		
Digoxin	Amoxycillin	Frusemide
Phenoxymethyl penicillin	Hydralazine	Potassium chloride
	Captopril	Spironolactone[2]
	Chlorothiazide[1]	Propranolol[2]
Epilepsy		
Sodium valproate syrup	Diazepam	Sodium valproate liquid
Phenytoin	Nitrazepam	Carbamazepine
Ethosuximide		Phenobarbitone
Promethazine		
Cystic Fibrosis		
Flucloxacillin	Amoxycillin	Amoxycillin & clavulanic acid
Ampicillin	Erythromycin	Ketovite
Cefaclor	Naldixic acid	Trimethoprin
Cephalexin	Cotrimoxazole	
Colistin	Multivitamin preparation	

Notes: [1] Extemporaneously prepared as a liquid oral preparation in hospital pharmacies.
[2] Available as a sugar-free liquid oral preparation from manufacturers on a "named patient" basis.

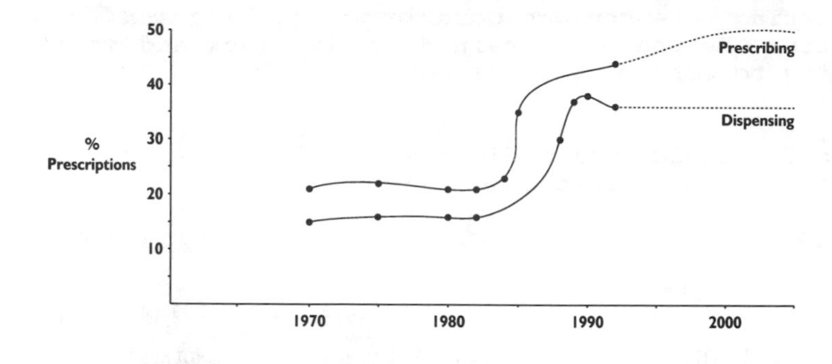

Figure 6. Changes in generic prescribing and dispensing 1970-1992 and predictions for the Year 2000, showing % of prescriptions in Great Britain prescribed as generic drugs, compared with the lower % of prescriptions dispensed as generic drugs (Source; Dhalla 1993[17]).

6 LIQUID ORAL MEDICINES CONTAINING SUGAR HAVE ASSOCIATED PROBLEMS: WHAT ARE THE OPTIONS FOR CHANGE?

Solid dosage forms

The use of solid dosage forms should be encouraged whenever possible; This is not always possible in young children but dosage forms such as sugar-free chewable tablets are now appearing on the market.

Sugar-free equivalent liquids

If solid dosage forms are not available or not well tolerated especially by younger children then sugar-free equivalent liquids should be actively promoted when they are necessary for chronic use.

Manufacturers

Manufacturers need to be aware of which drugs are likely to be required for long-term use, and generic manufacturers need to produce sugar-free liquid alternatives for long-term use.

Prescribers and dispensers

The majority of prescriptions for long-term use in children are provided by General Medical Practitioners, usually following the advice of hospital physicians.

Information and support regarding the use of sugar-free alternatives should be more forthcoming from the medical, dental and pharmacy professions. If a doctor is involved in the repeat prescribing of a liquid oral medicine long-term then he or she should be advised to specify "SF" or "sugar-free" on the prescription, since at present a pharmacist may contravene his terms of service if he supplies a sugar-free product when the prescription does not specify sugar-free.

Table 8. The sweetening agent contained in the 381 liquid oral medicines prescribed long-term according to the specificity of prescribing

	Number prescribed by generic name (%)	Number prescribed by proprietary name (%)	Total
Sugar-based	62 (39%)	60 (27%)	122
Sugar-free	32 (20%)	111 (50%)	143
Variable	66 (41%)	50 (23%)	116
TOTAL	160 (100%)	221 (100%)	381

The Department of Health

The Department of Health could tighten guidelines regarding the manufacture of generic medicines with regard to sugar content, and revoke the present EC export subsidies and production refunds which apply to finished medicinal products containing sucrose[18]. As generic prescribing and dispensing increase further towards the year 2000, generic manufacturers should provide a suitable range of sugar-free liquid generic medicines, which as well as being safe, efficient, and cost-effective, avoid risk to the dental health of those individuals who need to take medication long-term.

References

1. G. Nikiforuk, 'Understanding dental caries: Etiology and mechanisms', Karger, Basel, 1985, Chapter 1.

2. A. Thylstrup and O. Fejerskov, 'Textbook of Cariology', Munksgaard, Copenhagen, 1986.

3. E. Newbrun, 'Cariology', Williams and Wilkins, Baltimore, 1988, 2nd Edn. pp 102, 128.

4. A.J. Rugg-Gunn and W.M. Edgar, Community Dent. Health, 1984, 1 85-92.

5. A.J. Rugg-Gunn, 'Nutrition and dental health', Oxford University Press, Oxford, 1993.

6. Department of Health. Committee on Medical Aspects of Food Policy. Report of Panel on Dietary Sugars. No. 37. Dietary Sugars and Human Disease, HMSO, London, 1989, pp 16-20, 41-42.

7. Department of Health, PL/CDO (92), 1, HMSO, London, 1992

8. J.C. Boylan in 'The theory and practice of Industrial Pharmacy', Lea and Febiger, Philadelphia, 1986, 3rd Edn. (Ed. Lachmann, L., Leibermann, H.A. and Kanig, J.L.) pp 457-478.

9. A. Maguire, PhD Thesis, University of Newcastle upon Tyne, 1992

10. A. Maguire and A.J. Rugg-Gunn, Community Dent. Health, 1993, In press.

11. K. Palmer, The Practitioner, 1989, 233, 1464.

12. Office of Health Economics, 'Compendium of Health Statistics', HMSO, London, 1992, 8th Edn, 49-87.

13. I.F. Roberts and G.J. Roberts, J. dent. Child., 1981, 48, 346-351.

14. D.J. Kenny and P. Somaya, J. Can. dent. Ass., 1989, 55, 43-6.

15. A. Maguire and A.J. Rugg-Gunn, J. Dent. Res., 1992, 71, 749.

16. Department of Health, Statistics and Research division, Prescribing data, HMSO, London, 1988.

17. M. Dhalla, Pharm. J., 1993, 250, 238-41.

18. S.W. Bond and C.D. Fields, in 'Sugarless - The Way Forward', ed. A.J. Rugg-Gunn, Elsevier Applied Science, London, 1991, 154-162.

Labelling Requirements for Foodstuffs

Susannah L. Johnson

PRINCIPAL SCIENTIST – FOOD LEGISLATION DEPARTMENT, LEATHERHEAD FOOD RESEARCH ASSOCIATION, RANDALLS ROAD, LEATHERHEAD, SURREY KT22 7RY, UK

1 INTRODUCTION

Labelling is the key to informed consumer choice in the purchase of food and is, therefore, of prime importance. It is also a main element of the EC's harmonisation programme, which concentrates on a more informative system of food labelling backed up by essential food safety and hygiene measures; in addition, informative labelling is considered to be a better way to tell the consumer about the composition of foodstuffs.

In the UK there are labelling rules for sweeteners sold as such and for foodstuffs containing sweeteners. Owing to the UK's responsibility with respect to membership of the EC, many of the labelling rules are based on EC provisions. Future labelling provisions will also be heavily dictated by EC provisions.

2 LABELLING REQUIREMENTS FOR SWEETENERS SOLD AS SUCH

In regard to the labelling of sweeteners when sold as such, the required labelling information depends on whether the sale is a consumer or business sale and, if the sweetener is mixed with other substances, the nature of the substances.

For instance, sweeteners and sweeteners that contain "supplementary materials" (i.e. materials used to facilitate the storage, sale, standardisation, dilution or dissolution of the additive) must be labelled in accordance with the provisions of the Food Additives Labelling Regulations 1992[1] for both business and consumer sales.

In regard to sweeteners that contain substances that are not included in the narrower definition of "supplementary material":-

i) if they are not ready for delivery to the

ultimate consumer or to a catering establishment they must be labelled in accordance with the provisions of the Sweeteners in Food Regulations 1983[2], as amended.

ii) if they are ready for delivery to the ultimate consumer (that is, for retail sale) or to a catering establishment, they must be labelled in accordance with the provisions of the Food Labelling Regulations 1984[3], as amended.

In most cases the labelling of sweeteners, including table-top sweeteners, falls within the scope of the Food Additives Labelling Regulations.

A popular claim that is made on the labels of table-top sweeteners is that they are "low-calorie"; controls for such slimming claims are contained in the Food Labelling Regulations 1984 and the conditions imposed for such claims must be complied with.

In regard to net weight indications, these are controlled by a separate set of regulations. Other regulations that are applicable in certain instances are the Food (Lot Marking) Regulations 1992[4].

The main labelling requirements are detailed below.

Food Additives Labelling Regulations

In regard to the Food Additives Labelling Regulations, these Regulations implement Articles 7 and 8 of the Food Additives Framework Directive 89/107/EEC[5], which relate to labelling, and contain the following labelling requirements for consumer and business sales of sweeteners.

Labelling Requirements for Business Sales of Sweeteners. The container must bear the information required by either (a) or (b) below.

a) i) For each additive, where details of the name and EEC number are contained in the regulations, the correct name and number, and, for any other additive, a description that is sufficiently precise to enable it to be distinguished from any additive with which it could be confused. Where more than one additive is present, the information must be given in descending order of the proportion by weight of the additives in the container.

ii) Where any supplementary material is present, an indication of each component of the supplementary material in descending order (if more than one component) of the proportion by weight of the components.

iii) The statement "for use in food" or "restricted use in food", or a more specific reference to its intended food use.

iv) Any special storage conditions.

v) Any special conditions of use.

vi) Any instructions for use where it would be difficult to make appropriate use of the additives in the absence of such instructions.

vii) An identifying batch or lot mark.

viii) The name or business name and address of the manufacturer, or packer or EC seller.

ix) The percentage of each component of the food additives where it is prohibited to exceed a specified quantity or proportion of that component in any described food in which that component is used. Alternatively, sufficient information must be given to enable the purchaser to decide whether, and to what level, he could use such food additives in food sold by him without contravening any prohibition provision.

b) The information required by a) (i), (iii), (iv) and (v) above, and, in a conspicuous place, the words "intended for manufacture of foodstuffs and not for retail sale". Trade documents relating to the relevant consignment must be supplied to the purchaser with or in advance of the consignment and contain the information required by the remainder of (a) above.

The information and, where relevant, the words must be conspicuous, clearly legible and indelible.

Labelling Requirements for Consumer Sales of Sweeteners. The following information must be stated:

i) The name of the product. This is a description of the additives that are specified in the regulations and the EEC number relevant to each additive, or, where no description is specified in the regulations or where additives comprised in the description have no EEC number, a description that is sufficiently precise to enable the product to be distinguished from any product with which it could be confused.

ii) The information specified under Labelling Requirements for Business Sales of Sweeteners a) (i-viii) above.

iii) The minimum durability of the product, in the prescribed format.

The information required above must be conspicuous, clearly legible and indelible.

Sweeteners in Food Regulations

These Regulations require that sweeteners must be marked or labelled with the following:-

i) The name of the permitted sweetener.

ii) The serial number, if any.

iii) The words "for foodstuffs (restricted use)".

iv) The name or business name and an address or registered office of the manufacturer or packer, or of an EC seller.

v) For sorbitol syrup (E420) containing, after hydrolysis, a level of total sugars exceeding 1%, the words "contains after hydrolysis a level of total sugars of more than 1%" or similar words.

vi) Where a permitted sweetener is mixed with any other substance (including another permitted sweetener), the name of every other substance in the mixture.

Requirements relating to manner of marking the information are also prescribed.

Food Labelling Regulations

Part III of these Regulations applies to food that is ready for delivery to the ultimate consumer or to a catering establishment. This part of the Regulations prescribes requirements relating to the following indications on the labels of prepacked foods:-

a) The product name.
b) A list of ingredients (not required for single-ingredient foods).
c) An indication of the appropriate durability.
d) Any special storage conditions or conditions of use.
e) The name or business name and an address or registered office of the manufacturer or packer, or of an EC seller.
f) Particulars of the place of origin if failure to give such particulars might mislead a purchaser to a material degree as to the true origin of the food.
g) Instructions for use, if necessary.

Detailed requirements relating to the above-mentioned indications and manner of marking the information are prescribed.

Part IV of the Food Labelling Regulations imposes controls for the making of certain claims, including slimming claims such as "low calorie"; one condition is that the foodstuff must be labelled with nutrition information (which should be given in accordance with the provisions of the EC Nutrition Labelling Directive 90/496/EEC[6] - refer to comments below).

Sweeteners Proposal

The EC proposal for a Council Directive on sweeteners for use in foodstuffs[7] includes proposed labelling provisions for consumer sales of table-top sweeteners. It is proposed that the sales description of a table-top sweetener must include the term "...-based table-top sweetener", using the name(s) of the sweetening substance(s) used in its/their manufacture. In addition, the proposal requires that a table-top sweetener containing polyols and/or aspartame must bear the following warnings:

- polyols: "excessive consumption may induce laxative effects"
- aspartame: "contains a source of phenylalanine".

This means that a small pack of tablet table-top sweeteners containing, for example, aspartame and maltitol, would have to be labelled with the sales description "aspartame and maltitol based table-top sweetener". The sweeteners would also have to be identified in the list of ingredients in accordance with the appropriate rules and both of the above-mentioned warning statements would have to be given as well.

The proposal also provides for rules to be drawn up concerning: the details that must appear on the labelling of foodstuffs containing sweeteners in order to make their presence clear; and warnings concerning the presence of certain sweeteners in foodstuffs.

3 LABELLING REQUIREMENTS FOR FOODSTUFFS CONTAINING SWEETENERS

In the UK, the labelling of sweeteners in foodstuffs is covered by the Food Labelling Regulations 1984, as amended. Owing to the UK's responsibility with respect to membership of the EC, the Food Labelling Regulations implement the provisions of the EC Labelling Directive 79/112/EEC[8], as last amended by Directive 89/395/EEC. The Food Labelling Regulations state that additives must be declared in the list of ingredients by the category name (as prescribed) representing the principal function of the additive, followed by the additive's specific name or EC assigned number, if any, or both. The category name "artificial sweetener" is prescribed; hence, artificial sweeteners, such as aspartame and saccharin, must be declared in the list of ingredients by the category name "artificial sweetener" followed by the specific name of the sweetener - for example "artificial sweetener: saccharin". (At present these sweeteners do not have EC assigned numbers.) If there is no suitable category name, the additive must be declared in the list of ingredients by its specific name; hence, other sweeteners, such as xylitol or mannitol, must be declared in the list of ingredients by their specific names.

However, there have been discussions at EC level on a proposed amendment[9] to Annex II of the Labelling Directive (79/112/EEC), which lists the additive categories prescribed for the purposes of ingredients listing. It is proposed to harmonise these additive categories with those set out in the Additives Framework Directive (89/107/EEC); for instance, it is proposed to replace the category name "artificial sweetener" with "sweetener". Such a measure will, therefore, affect the current UK requirements concerning the labelling of sweeteners such as xylitol and mannitol.

Another issue of interest relates to warning statements on the labels of foodstuffs containing sweeteners. In the UK both the Food Advisory Committee (FAC) and the Committee on Toxicity of Chemicals in Food, Consumer Products and the Environment (COT) have recommended that all foodstuffs (including table-top sweeteners) containing aspartame should carry a clear warning label that the sweetener acts as a source of phenylalanine; such a label would assist those suffering from phenylketonuria (PKU) to avoid consuming these products. Future legislation in this area is likely as the EC sweeteners proposal covers this issue -refer to above comments. (There are also warning provisions for foods for diabetics.)

A major trend in recent years has been the increased consumer awareness of nutrition issues, and the relationship between diet and health. As a result of this greater awareness, a variety of products has been developed to meet consumer demands, with non-sugar sweeteners being used as sugar substitutes. Consequently, we have seen a number of nutrition claims on the labels of foods, including claims such as "sugar free", "low sugar", etc., which, at present, are not covered by specific legislation, although the general provisions of the Food Safety Act 1990 concerning misleading labelling are applicable. Following concern that the increasing use of such claims was misleading consumers, the UK Food Advisory Committee (FAC) made recommendations[10] concerning regulatory controls over the use of certain nutrition claims, including sugar claims. However, MAFF delayed action on these recommendations because proposals on the subject were expected from Brussels. EC proposals failed to materialise so MAFF issued draft Regulations to introduce legislative controls on nutrition claims as recommended by the FAC. The proposals were intended to influence the Commission's thinking in this area and encourage the Commission to come forward with proposals. The Commission imposed a 12-month freeze on the UK proposals, which ended in April 1993, to give the EC itself time to introduce its own proposals. It is unlikely that the UK proposals will progress any further as the Consumer Policy Services (CPS) has issued a working document on claims[11].

In regard to the FAC recommendations for nutrition claims, these specify the minimum or maximum levels of the nutrient that can be present before a claim can be made and require nutrition declarations so the purchaser knows how much of a particular nutrient he or she is getting. The nutrition declarations should be given in accordance with the provisions of the EC Nutrition Labelling Directive (90/496/EEC), the provisions of which are discussed below. In general, reduced claims are relative, e.g. foods may be described as "reduced-sugar" if the total sugars content of 100 g/ml of the food is maximum 75% of that of an equivalent food for which no such claim is made (this does not apply to reduced-sugar jam, for which there are separate provisions specified in the jam regulations). However, in order to make a "sugar-free" claim the total sugars content of the food must not exceed 0.2 g/100 g/ml.

In regard to the EC proposed provisions relating to claims, the CPS document includes general and specific requirements for claims, including requirements relating to health claims, claims relating to the low/weak or rich/high content of energy, certain nutrients and ingredients and claims relating to the reduced/increased and the absence/non-addition of certain substances.

For example, it is proposed in the CPS document that if a reduced-sugar claim is made for solid foods there must be at least a 25% reduction of the nutrient compared with a normal food and the reduction must be significant, that is, there must be more than 5 g/100 g reduction compared with a normal food; other conditions are also specified. The setting of an absolute limit (of 5 g) to denote a significant reduction will be, in general, a new concept in UK food legislation.

In addition, if a "no-sugar" claim is made for a solid food the sugars content must be less than 0.5 g/100 g (cf UK recommendations); other requirements are also specified.

The current CPS draft is complex in nature, and there have been numerous comments to this effect. However, it is understood that this draft has been revised and the revised version should be available shortly.

Another issue of relevance is nutrition labelling. As previously mentioned, when nutrition labelling is provided it should comply with the provisions of the EC Nutrition Labelling Directive (90/496/EEC); in the transitional period before the Directive is implemented into UK law, following agreement with LACOTS (Local Authorities Coordinating Body on Food and Trading Standards), MAFF has advised that the provisions of the Directive should be followed fully.

Under the Directive nutrition labelling is voluntary, unless a nutrition claim, for example "low-calorie", "sugar-free" is made, in which case it is compulsory. Where nutrition labelling is provided, the information given must consist of that information prescribed in either Group 1 or Group 2.

Group 1 consists of energy, protein, carbohydrate and fat, and Group 2 consists of energy, protein, carbohydrate, sugars, fat, saturates, fibre and sodium (in the order given).

A further optional list of nutrients, which includes polyols, that may also be declared is given. Where a nutrition claim is made for sugars, saturates, fibre or sodium, the information given must consist of Group 2. Requirements relating to the derivation of declared values and manner of marking the information are also prescribed.

Revised proposals to implement the provisions of the Directive have been circulated for comment; the proposals were revised following the numerous comments received during the first consultation exercise. Subject to the outcome of the consultation exercise and Parliamentary procedures, MAFF hopes to bring the regulations into force as soon as possible after 1 October 1993 (which is the implementation date cited in the Directive), to permit formally trade in products that are labelled in accordance with their main provisions and to prohibit non-complying products; however, there are transitional arrangements with respect to compliance with the vitamin and mineral information.

4 CONCLUSION

Developments in labelling requirements will continue and these will be heavily dictated by EC measures. This, therefore, means that manufacturers, etc., must keep up-to-date with the impending changes, so that they can establish the practical and economic implications of proposals, make representations where necessary, and ensure that their products comply with all the necessary requirements.

Labelling will also continue to play a vital role in the informed consumer choice in the purchase of food, although a question often raised is how far does it have to go to meet this need?

REFERENCES

1. The Food Additives Labelling Regulations 1992 SI 1992 No. 1978.
2. The Sweeteners in Food Regulations 1983 SI 1983 No. 1211, as amended by SI 1988 No. 2112.
3. The Food Labelling Regulations 1984 SI 1984 No. 1305, as amended by SI 1989 No. 768, SI 1990 No. 2488 and SI 1990 No. 2489.
4. The Food (Lot Marking) Regulations 1992 SI 1992 No. 1357.
5. Council Directive 89/107/EEC of 21 December 1988 on the approximation of the laws of the Member States concerning food additives authorised for use in foodstuffs intended for human consumption. Off. J. European Communities, 1989, 32 (L40), 11.2.89, 27-33.
6. Council Directive 90/496/EEC of 24 September 1990 on nutrition labelling for foodstuffs. Off. J. European Communities, 1990, 33 (L276), 6.10.90, 40-4.
7. Proposal for a Council Directive on sweeteners for use in foodstuffs, COM(92) 255 final - SYN 423. Off. J. European Communities, 1992, 35 (C206), 13.8.92, 3-11.
8. Council Directive 79/112/EEC of 18 December 1978 on the approximation of the laws of the Member States relating to the labelling, presentation and advertising of foodstuffs. Off. J. European Communities, 1979, 22 (L33), 8.2.79, 1-14, as amended by:
   Council Directive 86/197/EEC of 26 May 1986. Off. J. European Communities, 1986, 29 (L144), 29.5.86, 38-9.
   Council Directive 89/395/EEC of 14 June 1989. Off. J. European Communities, 1989, 32 (L186), 30.6.89, 17-20.
9. Draft Commission Directive amending Directive 79/112/EEC on the approximation of the laws of the Member States relating to the labelling, presentation and advertising of foodstuffs. Document III/3630/91-EN-Rev.4.
10. Food Advisory Committee recommendations on nutrition claims in food labelling and advertising. May 1989.
11. Consumer Policy Services draft proposal for a Council Directive on the use of claims relating to foodstuffs. Document SPA/62/ORIG-Fr/Rev.2, 23.11.92.

Diet and Dental Problems in Children

A. S. Blinkhorn

DEPARTMENT OF ORAL HEALTH AND DEVELOPMENT,
UNIVERSITY DENTAL HOSPITAL, HIGHER CAMBRIDGE STREET,
MANCHESTER M15 6FH, UK

Over the past fifteen years epidemiologists have charted a pattern of decreasing caries incidence amongst children in the majority of industrialized societies[1,2]. Most authors suggest[3,4] that the widespread use of fluoride dentifrice is the main reason for the decline in dental caries; however other factors such as eating patterns, dental awareness and improved standards of oral hygiene have also been of some importance.

The positive change in children's oral health is a major health gain which has been achieved at little or no cost to either government or health service agencies. The improvement has essentially been led by the dentifrice manufacturers who have improved their products and persuaded the consumers to buy them.

The overall decline in caries rates must however be treated with some caution as mean reductions hide the fact that many children still suffer considerable dental health problems. The majority of caries is concentrated in children from lower socio-economic groups. Dental caries can be classed as a 'disease of deprivation'. Table 1 highlights the magnitude of the differences between upper and lower social groups in Glasgow, Scotland[5]. It can be seen that 67 per cent of five years olds from social classes I, II and III NM are caries free compared with 17 per cent in the other social grouping. Such a difference is both socially and politically unacceptable, given that the children live in relatively close proximity to one another. A large difference in the proportion of caries free children was also found to be related to social class in 12 year old Glaswegians. Table 1 shows that 33 per cent of upper social grouping were caries free compared with 9 per cent in the lower social grouping.

The other cause for concern is that Downer[6] has suggested that caries rates in five year olds in England and Wales have stabilised and may indeed be rising[7]. This could be a consequence of the recession. Unemployment will change both toothpaste buying habits and dietary behaviour. Thus the dental health problems of children have a social class component.

Table 1. Comparison of mean dmft* and DMFT+ for five and 12 year old children in Glasgow according to social class

5 year olds

Social Class	dmft	(S.D.)	% caries free
I, II & III NM	2.67	(3.97)	67
IIIM, IV & V	5.41	(4.30)	17

12 year olds

Social Class	DMFT	(S.D.)	% caries free
I, II & III NM	2.14	(2.28)	33
IIIM, IV & V	3.79	(2.57)	9

* decayed, missing and filled primary (baby) teeth
+ Decayed, Missing and Filled Permanent (Adult) Teeth

Social class is an important variable when discussing norms of behaviour as there is often an 'information gap' between different social groupings in terms of health behaviour[8]. This disparity is highlighted in a study which recorded the dietary habits of nursery school children living in deprived and non-deprived areas of Scotland[9]. The area in which a family lived was a major factor influencing the pattern of sugar consumption. Table 2 shows that the majority (75 per cent) of mothers from the deprived areas gave their children sweets after nursery school; this finding contrasts with the non-deprived area where only 32 per cent of the mothers stated that sweets were given at this time. 94 per cent of the mothers from the deprived areas believed it was normal practice to give children sweets when they came out of nursery school. In the non-deprived areas no such social expectation was evident as only 38 per cent expected other mothers to give their children sweets at the end of the nursery school session.

Table 2. Sweets usually given to children after nursery school according to the area of residence

Sweets given	Deprived Area	Non-deprived Area
	%	%
Yes	75	32
No	25	68

Mothers whose children went to nursery schools in the non-deprived areas were more likely to prefer savoury foods or fresh fruit; 42 per cent compared with only 10 per cent of children from the deprived areas. Interestingly the reasons for giving sweets differed between the two groups (Table 3). Mothers from the non-deprived areas placed more emphasis on reward as a reason for giving sweets, whereas the other group of mothers gave sweets to keep the child quiet or because the child demanded sweets.

Considerable differences were also found in the timing of the children's reported sweet consumption (Table 4). Mothers from deprived areas were more likely to give their children sweets after nursery and allow them sweets anytime during the day. The sweet consumption of the children from the non-deprived areas was far more restrictive. Striking differences were also noted when the mothers were asked to report on the money they usually spent on sweets each week (Table 5). The mothers from the prosperous area were more frugal than their counterparts in the deprived area, a finding supported by Dummer et al[10] who reported that the amount of money spent on sweets per week had a clear association with caries experience.

The reported differences in the pattern of children's sweet consumption was reflected in the dental data. Table 6 shows that the mean dmft scores for the primary dentition in the deprived areas was 3.20 compared with 0.88 for the non-deprived areas: a difference of over 70 per cent. Dental caries is clearly a major problem for the socially disadvantaged children included in this study, as the decayed (d) component differed by 74 percent between the two groups (Table 6).

Table 3. Mothers reported reasons for giving sweets according to area of residence

Reasons	Deprived Area	Non-deprived Area
	%	%
A reward	19	61
Keep child quiet	24	10
Child demands sweets	57	29

Table 4. The usual time for most sweet consumption according to area of residence

Reasons	**Deprived Area**	**Non-deprived Area**
	%	%
After nursery	45	29
Weekends	6	42
Any time	36	3
After meals	13	26

Table 5. Money spent on sweets per week according to area of residence

Money Spent	**Deprived Area**	**Non-deprived Area**
	%	%
Up to 50p	31	84
50p to £1	37	9
£1 +	31	7

Table 6. Mean dmft scores according to area of residence

	Deprived Area	**Non-deprived Area**
dmft	3.20	0.88
d	2.80	0.74
m	0.30	0.02
f	0.19	0.12

Baric et al[11] reported that in their study of mothers with young children in a deprived area of England, the majority believed that sweet foods in general were a highly palatable foodstuff which children should enjoy and confectionery was clearly perceived as something special. Indeed sweets were closely linked with affection and not giving them was associated with deprivation. It was the norm to give sweet foods. An interesting finding was the behaviour of grandparents. The older generation irrespective of social class had a less constrained approach to sugary foods, especially sweets.

A study of 14-16 year olds suffering from adolescent rampant caries highlights the importance of social background on dental health and the influence of unrestricted sugar consumption on caries prevalence. Dietary data on one hundred adolescents who had been referred to the Glasgow Dental Hospital over a five year period because of high caries rates were examined retrospectively[12]. Rampant caries for the purposes of this study was defined as children with eight or more lesions or restorations in the upper and lower permanent incisors and canines. All the subjects had completed a diet diary soon after their initial consultation, as well as a questionnaire on dietary attitudes and dental visiting behaviour. These were examined for trends in dietary behaviour, and attitudes to dental care. The children's postal addresses were used to give some guidance on the potential social background of each child. The results indicated that 87 of the subjects were from working class areas, only 18 of whom were female. Interestingly the majority of adolescents with a middle class background were female, 9 out of the 13 subjects.

Analysis of the dietary and questionnaire data revealed a number of common factors which linked the group together.

1. The majority (79) received little cooked food at home, hot food was usually obtained from a local 'take-away' shop; chips (french fries) with a bread roll being the most popular choice.
2. This group of adolescents did not have set mealtimes, their eating behaviour could best be described by the term 'grazing'. Snacks, usually proprietary confectionary, high in sugar, together with sugary drinks were taken frequently, particularly at the following times: 8.15-9.30, 10.30-11.00, 12.30-14.00 and 15.45-23.00 hours. Snacking was an activity which was most popular after school and peaked in the early evening from 18.30 to 22.30.
3. Eighty-five of the children were extremely frightened of dental treatment and had only presented for care when they felt their appearance was threatened. The remainder were not frightened but complained that dentists did not understand them; they felt as though they were being blamed for the state of their teeth for no good reason.
4. Oral awareness was of a low order as all the subjects had poor plaque control and generalised gingivitis.
5. When asked about eating habits in general, most of the subjects (96) felt that their eating habits were similar to their friends. They did not feel their behaviour was unusual.

This group of adolescent subjects represents the extreme of a continuum of behaviour that to a lesser degree is common amongst many young people, namely 'grazing' on snacks rather than eating at regular intervals as recommended by the Health Education Authority[13]. Why these youngsters should have much higher caries rates than their peers is most interesting, as in behavioural terms they thought that their diet was similar to that of their friends.

The problem faced by those seeking to promote better dental health through dietary control is the fact that different groups in society require different messages presented in ways that are non-threatening[14].

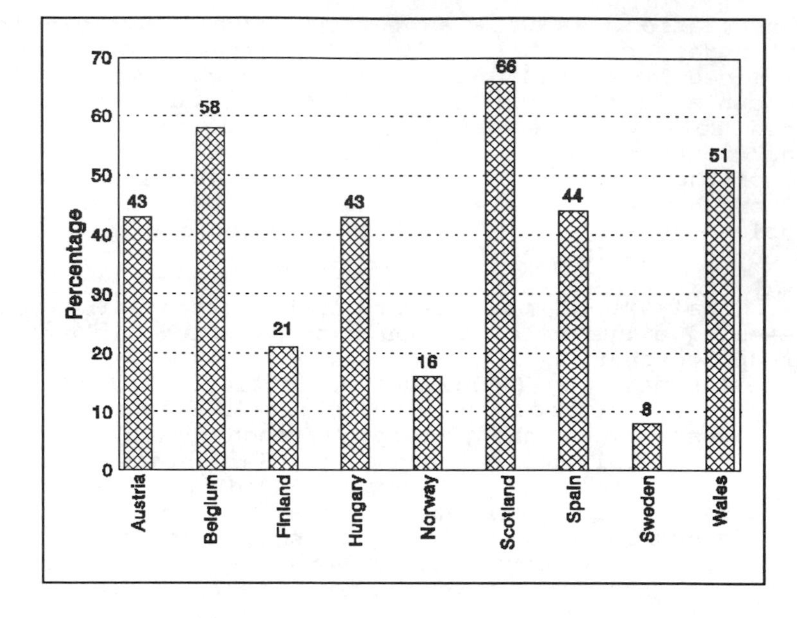

Fig. 1 Daily consumption of sweets among 7 year olds in some European countries (Honkala et al 1990,[15])

Many middle class families have curtailed sugar consumption but their other health behaviours such as use of fluoride toothpaste will also exert a major influence on caries control.

For the deprived families the dental future is less favourable; sugar seems an all pervasive feature for both the young child and the adolescent. Frequent use of sugary foods is a behaviour supported by social expectations. In most European countries[15] approximately half of the children consume sweets and soft drinks on a regular basis (Fig. 1).

If a general sociological model is applied to dental health education the importance of socialisation becomes apparent[16]. Primary socialization is the time when the earliest childhood routines are acquired. This coincides with the establishment of dietary habits which will greatly influence future patterns of sugar consumption. In addition to the acquisition of the routines and habits during primary socialization, the period of secondary or formal socialization is important. This is when knowledge is acquired which is necessary for a rational explanation of behaviour and which serves as a reinforcement for the maintenance of behaviour[17]. Thus those interested in promoting healthy eating habits must take account of the socialization process when preparing advice. For families with young children the advice agenda could concentrate on establishing healthy habits as primary socialization is a time for forming behaviour patterns. During the secondary socialization phase education should attempt to modify existing habits as changing ingrained habits completely is difficult and time consuming. 'Modification' is the key activity.

The dietary habits of children and their relationship to dental caries is an area of potential conflict between the food industry and many health professionals.

Parents want to do 'what is best' for their children in terms of a healthy diet but are often unsure about what is good dietary practice. Many people do not understand what is meant by fat, protein, carbohydrate, nor do they grasp the difference between carbohydrate and calories. Furthermore, the advice offered by many health professionals is too general and may lead to misinterpretation. Information should be tailored to individual needs and take account of the particular social mores of different population subgroups.

It is generally accepted that sugar is the most important aetiological agent influencing the development of dental caries. Such a statement however arouses the ire of the sugar manufacturers who point out that dental caries is a multifactorial disease, and it is the frequency rather than the amount of sugar which is important. On the first point most authorities agree that sugar causes dental caries. The frequency versus the amount debate is one that really confuses the public and prejudices the success of dietary advice programmes. Although frequency is important, the number of intakes per day is closely related to the amount of sugar consumed, and this point has been used by the sugar industry to confuse people.

The type of dietary advice that should be offered to parents is further complicated by the dramatic changes in the prevalence of dental caries and the segmentation of the population into those at high and low risk to caries. Clearly low risk families only require general dietary advice whereas high risk groups will need detailed counselling.

Also at one time dental decay was so prevalent that there was little interest in distinguishing the different 'attack sites' on the teeth. However with a decline in disease levels dental scientists now breakout their data by two main surface types, namely the occlusal (biting) surfaces which are characterised by having a complex system of fissures, and smooth surfaces which make up the rest of the crown of the tooth. The fissure systems on the occlusal surfaces are the most susceptible to dental decay. This can be demonstrated by examining American data (Fig. 2)

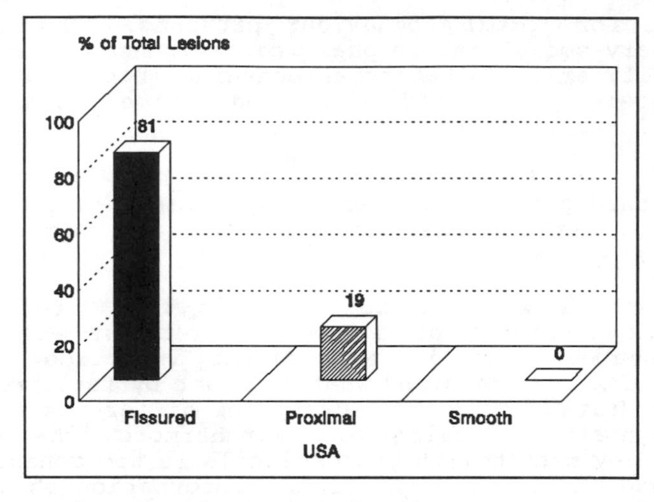

Fig. 2. Distribution of caries increments by pit-and-fissure, proximal, and free smooth surface for children in diet studies in Michigan, USA. Data from Burt et al[18].

published by Burt[18]. He demonstrated the existence of a group of susceptible children who suffered a much higher level dental decay in the occlusal surfaces of their permanent teeth. This is because fluoride in toothpaste or water is most effective in preventing dental decay on smooth surfaces and exerts only a minor effect on the fissures of the occlusal surfaces.

Thus there are two inter-related problems that researchers have to consider when trying to improve children's dental health. First, the majority of dental disease is concentrated in a group of susceptible children, and second the disease originates in the fissures which are resistant to one of the most powerful preventive agents - fluoride.

This poses difficulties for public health dentists and health promotion officers trying to design and target preventive programmes. Dental epidemiologists report that the susceptible children tend to live in poorer areas; hence dental decay in children is associated with social deprivation. It is well known that deprivation limits the success of health promotion as parents have so many problems to contend with that many aspects of health are given a low priority.

Indeed behavioural scientists have pointed out that illness is not solely a biological phenomenon but occurs within a social context[19]. Social factors may initiate or promote a disease process and can operate so as to affect the efficacy of oral health promotion. A child's participation in and acceptance of oral health promotion programme is not a random event but is structured by a variety of demographic, organizational and social factors influencing parental action and interest. Individuals in lower socio-economic groups are exposed to more unhealthy lifestyles and this will result in health inequality.

Many obstacles to oral health may have their roots in sociocultural norms of the society in which people live, and as such there may be differing responses to oral health advice. Social factors may be indirect measures of other undetected factors[20]. For example, ethnic status may indicate that families will have particular dietary habits but these may overlap with language barriers, low income, inadequate housing and poor mobility. Lifestyle will therefore have a profound effect on oral health, as the social support networks people establish will influence diet, oral self care practices and general health beliefs. Lifestyle relates very closely to social networks established by families. Historically the family has been the major factor in developing health care practices. However, it is clear that in many industrialized societies there have been considerable changes in the idea of 'what constitutes a family'. For many children fathers are not party to their upbringing and single mothers have to fight against fiscal and social pressures which were more appropriate to a family lifestyle of twenty years ago. Single parent families may become isolated from the social network and advice or actual care may not be easy to obtain. The extended family used to have a critical role in encouraging and sustaining health care practices.

It is clear that children are not free agents and there is evidence that a child's oral health status is closely correlated with dental health status of the mother[21]. The ability of a child to participate in improving health is limited by parental willingness to consider health as an important priority, for example, when purchasing food. A child may receive information on oral health threats through a school based education programme but may not receive support at home if parents have different values or lack the resources to comply with the suggested information. For example, there have been attempts to deny children all sugar products, but this would mean changes in the social norms whereby sweets are equated with reward and present giving. There is also the problem of over-estimating the priority given to oral health issues. Many dedicated health workers giving advice on the dietary control of caries[22] have been accused of 'victim blaming' or 'healthism'. As a solution to such a claim it has been suggested that sweets should be given just once a day or once a week on a special day. This is classed as modification of behaviour rather than trying for a complete change.

So the education of a child may be compromised on the one hand by health professionals offering unrealistic advice and on the other by the refusal of parents to change behaviour because of cultural or fiscal barriers. The child is caught between the clash of two authoritarian systems.

As a child matures into adolescence so the social pressures change and the sources of influence alter. The importance of the family wanes and peer group pressure becomes the focus of influence and concern[23]. Advertizing feeds on this environment of uncertainty and insecurity by fostering the need for conformity[24]. Success and popularity depends on the 'right' fashion accessories, chewing gum, and confectionery. Health promotion planners have realised that negative advice does not fit easily into adolescent culture hence the switch to positive health education. A healthy lifestyle is promoted as the 'natural thing to do'. Unfortunately this type of effort is compromised by a lack of resources when compared with the advertizing campaigns undertaken by commercial companies.

Thus making healthier choices 'easier choices' may be compromised by the inability of the health professionals to compete with the lifestyle propagated by media based advertizing campaigns produced by commercial companies. Indeed health promotion is now centering on 'empowerment', whereby individuals are shown that they have a measure of personal control. The issue of self esteem and personal responsibility for health is seen as one way of countering the broad based lifestyle campaigns of commercial companies.

Advertizing any products for human use or consumption requires some ethical consideration. A considerable number of sugar products, for example, have been marketed continuously for many years and manufacturers often claim that they are seeking to maintain market share rather than expand a particular market. However when new products are produced there is a clear ethical requirement that they should not cause or exacerbate health problems.

Rampant dental caries[25] is a problem for young children who have been allowed to consume sugary drinks on a continuous basis. Night time consumption of sugary drinks is particularly harmful because the flow of saliva is reduced and the oral environment becomes highly acidic resulting in the rapid demineralisation of dental enamel. Rampant dental caries causes considerable pain to the children and much anguish to parents, who are usually unaware of the dental dangers of sugar drinks. Despite the well proven dangers of sugar drinks some manufacturers produced sugary drinks specifically for use at bed time. Clearly here is a case where any accord between health professionals and manufacturers is impossible.

CONCLUSION

Oral health has improved dramatically in the industrialised nations but dental caries remains a major problem for many children. The prevalence of dental caries is related to the amount and frequency of sugar consumption and as such further reductions in caries are dependent on changing the dietary habits of a substantial number of families.

REFERENCES

1. M.C. Downer, Health Trends, 1989, 21, 7.
2. T.M. Marthaler, Caries Res., 1990, 24, 381.
3. K.G. Konig, Caries Res., 1993, 27, 23.
4. G. Rolla, B. Ogaard, Strategy for dental caries prevention in European countries according to their laws and regulations. Eds. R.M. Frank, S. O'Hickey, Oxford, UK, 1987, p.223.
5. A.S. Blinkhorn, J. Cummins, A.S. MacMillan, G. O'Mailley, Br. Dent. J., 1986, 160, 287.
6. M.C. Downer, Caries Res., 1992, 26, 466.
7. R.D. Holt, Br. Dent. J., 1990, 18, 296.
8. P.E. Petersen, Community Dent. Oral Epidemiol., 1990, 18, 153.
9. A.S. Blinkhorn, Br. Dent. J., 1982, 152, 227.
10. P.M.H. Dummer, S.J. Oliver, R. Hicks, et al, J. Dent., 1990, 18, 37.
11. L. Baric, A.S. Blinkhorn, C. MacAurthur, Health Educ. J., 1974, 33, 79.
12. A.S. Blinkhorn, J. Inst. Health Educ., 1989, 27, 179.
13. R.S. Levine (ed), Scientific Basis of Dental Health Education - A Policy Document, Health Education Authority, London, 1985.
14. A.S. Blinkhorn, G.B. Hastings, D.S. Leathar, Br. Dent. J., 1983, 155, 311.
15. E. Honkala, L. Kannas, J. Rise, Int. Dent. J., 1990, 40, 211.
16. A.S. Blinkhorn, Int. Dent. J., 1981, 31, 14.
17. M. Craft, Int. Dent. J., 1984, 34, 204.
18. B.A. Burt, S.A. Ecklund, K.J. Morgan, et al, J. Dent. Res., 1988, 67, 1422.
19. H.C. Gift, Current Opinion in Dentistry, 1991, 1, 337.
20. M. Blaxter, Social Science and Medicine, 1983, 17, 1139.
21. R. Nowjack-Raymer, H.C. Gift, J. Publ. Health Dent., 1990, 50, 370.
22. E.J. Kay, A.S. Blinkhorn, J. Hlth. Educ. Res. Theory & Practice, 1986, 1, 307.
23. K. Hurrelmann, Human Development and Health, Springer-Verlag, Berlin, 1989.

24. C. Wright, Oral Health Promotion, L. Schou, A.S. Blinkhorn (eds), Oxford, UK, 1993, p.149.
25. L.W. Ripa, Paediatric Dentistry, 1988, 10, 268.

Diet and Dental Problems in Adults

A. W. G. Walls

DEPARTMENT OF RESTORATIVE DENTISTRY, THE DENTAL SCHOOL, UNIVERSITY OF NEWCASTLE UPON TYNE, FRAMLINGTON PLACE, NEWCASTLE UPON TYNE NE2 4BW, UK

1 INTRODUCTION

The dental problems of older patients are somewhat different from the young. Dental caries increment is at its height immediately after the eruption of both the primary and secondary dentitions as the vulnerable tooth surfaces are attacked in a cariogenic environment. There is an argument that there may be a second episode of high risk from caries as root surfaces become exposed in the mouth through periodontal disease. In addition to dental caries, the two chronic oral problems (periodontal disease and tooth wear) progress throughout life and tend to manifest themselves in severe form in older subjects.

It can be argued that diet and nutrition have a significant influence in all three of these oral problems, and it is this relationship that this chapter will address.

2 AN AGEING POPULATION

There have been, and will continue to be marked demographic changes in the population of the world. These changes comprise population growth globally, with an increase in young people in the third world and a gradual ageing of the population in the first. In Europe and America this ageing of the population is associated with an alteration in dental status, with an increasing number of old people retaining some of their natural teeth. Population projections over the next 30 to 40 years suggest that, in the over 75s in the UK, the rate of edentulousness will fall from 80 per cent to 20 per cent.[1] Over the same time period the population in that age-group will grow from 3.5 to 5.5 millions.[2] If it is assumed that the dentate over 75s retain at least as many teeth in 40 years time as they do now, then this change will result in an increase in the number of teeth in this age-group from in the region of 4.2 million to 61.6 million. This pattern of change is at its most dramatic in the UK, but as can be seen in Table 1 it is also occurring elsewhere in the western world.

Despite these improvements in dental status of older subjects there is an inevitable decline in numbers of teeth with increasing age. Such decline will have an influence on masticatory efficiency, at least at a subjective level,[3] as will the reduction in lean muscle mass and consequently biting force seen in older subjects.[4,5]

Table 1

Population projections for rates of edentulousness amongst 65+-year-olds in Europe[56]

Country	Most Recent Figure (%)	Projection for 2000 (%)
United Kingdom	67	49
Finland	65	30
Denmark	60	45
Germany (GDR)	58	20
Sweden	20	

Individuals with compromised dentitions overcome their masticatory problems by increasing the number of chewing strokes,[6,7] thus prolonging contact time in the mouth, by selecting foods which require less comminution prior to bolus formation[5] or by preparing their food in such a way that chewing is not required (e.g. dunking hard foods in liquid[8]).

3 DIET AND DENTAL DISEASE

Irrespective of the age of the individual, providing they have teeth, they will be subject to the two major disease processes which threaten the integrity of the dental arch, caries and periodontal disease. In addition, wear to the surfaces of teeth in function does occur. A degree of wear is a normal finding, however, wear can pose a problem in terms both of appearance and masticatory function if it occurs at an excessive rate. Diet can have an impact upon the severity of all three of these clinical problems.

Dental Caries

There are three forms of dental caries in the adult, new enamel caries, root surface caries and secondary decay around or beneath an existing, and often deficient, restoration. There is some information in the literature relating to both new coronal enamel and root surface lesions, but there is little information available about recurrent decay, either in terms of factors that influence the development of recurrent lesions, or in terms of the rate of development of such cavities. Anecdotal evidence would link an increased attack rate for all three conditions with an increased frequency of intake of dietary sugar, and also with poor standards of oral cleanliness.

It should be emphasised that dental caries is a disease of the environment. The environment can be modified in a number of ways in the adult to render a subject who had a low attack rate for dental caries susceptible to the disease.

The two major dietary intervention studies from Vipeholm[9] and Turku[10] have both demonstrated the relationship between sugar intake and the development of new carious lesions in adults. These results are confirmed by the observational data from subjects with hereditary fructose intolerance,[11,12] and in subjects whose diets have changed from low to high sugar contents.[13] These data relate to new smooth surface enamel lesions for the most part, although some root surface lesions were apparent in the Vipeholm study, especially in those subjects who ate sugar both at and between meals. There is little information available in the literature about the rate of development of enamel lesions in adults on a normal diet, although new coronal lesions do occur during longitudinal studies of dental health.[14]

The rate of development of new root surface lesions has been reported as 1.8 surfaces per hundred surfaces at risk per year in longitudinal studies of community dwelling older adults.[14] There is a suggestion from the Vipeholm study and from cross sectional analyses of diet and root surface caries that there is a positive relationship between frequency of dietary sugar intake and root caries.[15,16] However, it has also been noted that root surface lesions can develop in populations who have a high starch, low sugar diet[17,18] and root caries is a relatively common form of dental pathology in specimens from archaeological studies who lived at a time before the widespread availability of refined carbohydrate.

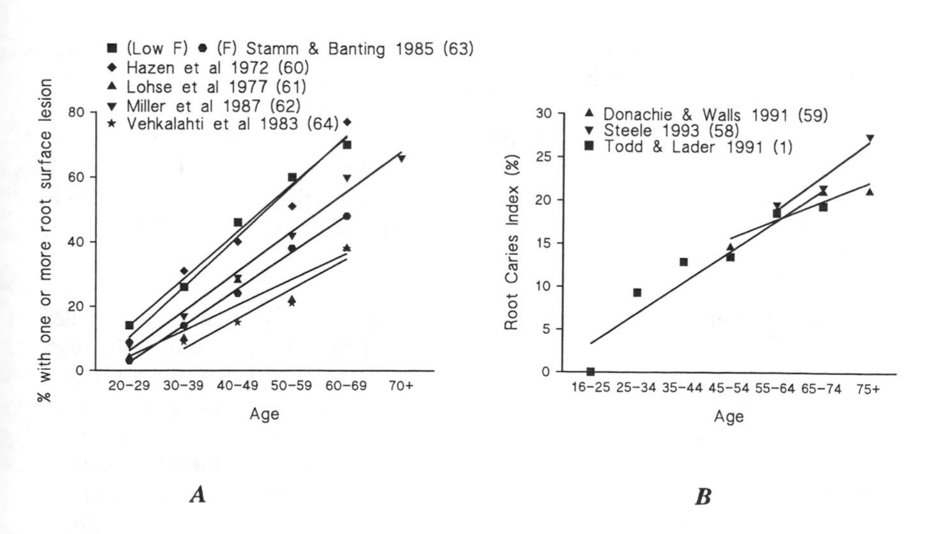

Figure 1 ***A.*** Increasing prevalence of root surface caries with increasing age from a wide variety of population studies. ***B.*** Increase in attack rate with increasing age using the Root Caries Index (RCI)[57]

It is interesting to note that both the prevalence and the attack rate for root surface caries increase with increasing age (Fig. 1). The increasing prevalence can be simply explained on the basis of an increase in the numbers of exposed root surfaces (which are thus at risk from developing surface decay) in an older population. However, over and above this anatomical change there appears to be an increase in attack rate for root surfaces from in the region of 18% of exposed surfaces being decayed or restored in the 60-year olds to 25-30% of such surfaces being decayed or restored in subjects over the age of 75. It may be that dietary factors play a part in this modification in disease profile. It has been noted already that older subjects, especially those with a reduced dentition, tend to use a softer diet, the remnants of which may be more adherent to tooth tissue. It has also been reported that older housewives purchase more packaged sugar per capita than their younger counterparts,[19] and that adults over the age of 60-65 derive more of their calorific intake from non-milk extrinsic sugars than do younger people.[20] These alterations in diet occur at a time when an older subject's manual dexterity and their tolerance of repeated, delicate neuromuscular tasks is reducing. This makes routine oral hygiene more difficult for an older person, especially when the intra-oral architecture that needs to be cleaned becomes more complicated simultaneously as a product of gingival recession. Thus a potential increase in intake of fermentable carbohydrate is associated with a diminishing ability to clean teeth, especially the root surfaces. This is a recipe for dental caries.

At the beginning of this section it was emphasised that dental caries is a disease of the environment, and that alteration in the environment can result in high caries activity. One such alteration is an increase in dietary sugar intake. However, there are other changes in the oral environment that can result in high caries activity even in subjects taking an acceptable diet.

The use of partial dentures, in association with inadequate standards of oral hygiene has been linked to increased dental disease, both coronal and root caries and periodontal destruction. The teeth most commonly effected are those adjacent to the edentulous areas, which are in close proximity to the denture base and its potential stagnation zones. Obviously, the use of partial dentures increases with age as teeth are extracted and some form of prosthetic replacement is required, although there is some debate in the literature concerning how many teeth we need with a *modern diet.*[21]

A second, and potentially major, risk factor is alteration in salivary flow in older subjects. Many old people report a sensation of dry mouth (xerostomia), although often this is not linked with flow rates that would be described as pathologically low.[22] However, there are a number of causes for profound reduction in salivary function which increase in prevalence with age. These include pathological destruction of gland tissues as an auto-immune phenomenon, radiation induced damage to the secretory cells and/or the use of drugs which inhibit salivary flow as a side effect of their pharmacological activity.

The functions of saliva are numerous, but perhaps the most important are the prevention of dental caries through a buffering effect on plaque acid derived from dietary sugars, by diluting and washing sugars out of the mouth through continuous

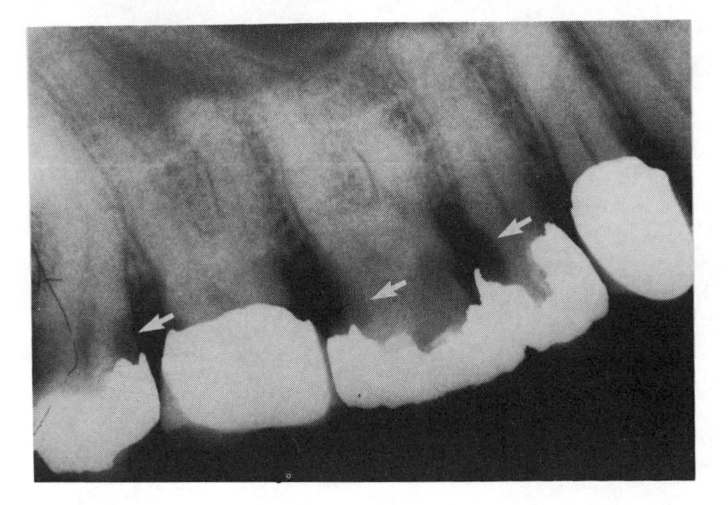

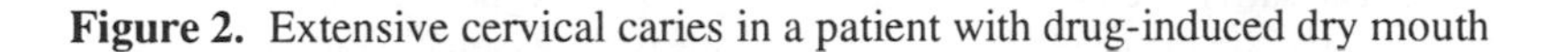

Figure 2. Extensive cervical caries in a patient with drug-induced dry mouth

salivary flow and through its activity as a vehicle for remineralisation of demineralised surfaces. The process of de and remineralisation occurs continually in the mouth as a result of ingestion of acids and sugars. If adequate saliva is not present then the balance is lost and caries results.

The pattern of decay seen in xerostomic subjects is pathognomic with widespread carious lesions on exposed dentine/cementum surfaces (Fig. 2). Such lesions not only occur on exposed root surfaces but also in any cups or fissures on tooth crowns and, in severe cases, on the incisal edges of teeth where dentine has been exposed as a result of functional wear. There is no reason to think that enamel surfaces would not decay also in this environment, it is simply that the dentine lesions progress with such rapidity that they demand urgent and active intervention. This takes the form of provision of any necessary restorations and active replacement of saliva and a targeted preventive strategy which would in turn help to minimise coronal decay.

One reason why there is a problem of xerostomia in older subjects is an age associated reduction in secretory tissue in the salivary glands.[23] These morphological changes do not impair salivary flow in the fit, healthy unmedicated individual, probably as a result of a small residual salivary reserve (Fig. 3).[24] However, even a relatively minor challenge, either pharmacological or from disease, can result in a significant reduction in salivary flow. Indeed, it has been demonstrated that such a reduction is greater in older subjects than in the young for a given bolus dose of an anticholinergic drug.[25] Older people tend to take more drugs than the young. In a recent study by Levy and co-workers[26] over 50 per cent of their sample over the age of 65 were taking at least one drug which had a potentially xerostomic side effect th 22 per cent taking more than one such agent. These drugs act on the salivary s in a number of ways, either by interfering with the secretory control mechans which are complex, or by altering systemic water balance. Whenever more an one xerostomic drug is being taken there is always the possibility for their mutual

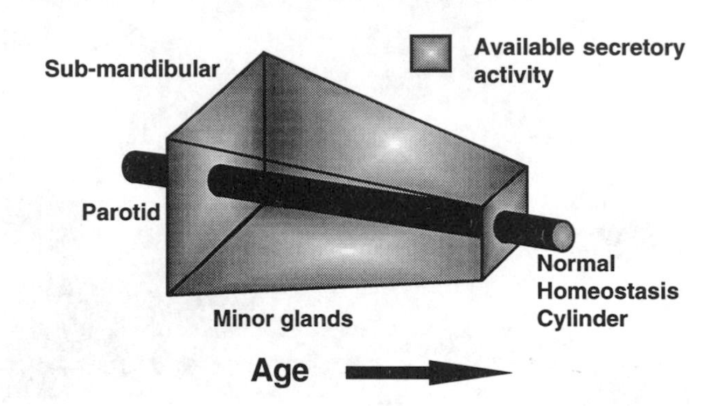

Figure 3. The salivary continuum, redrawn from Baum *et al.*[24] Secretory capacity is represented by the shaded cone, with a decrease in available output with increasing age. However, there is still adequate secretory capacity for *normal* flow in the healthy, unmedicated individual.

effects to potentiate each other producing a greater problem than either agent independently.

One clinical difficulty encountered when managing the xerostomic subject is alleviating their feeling of having a dry mouth. There are two possible approaches to this problem, salivary stimulation and salivary substitution. Obviously great care is needed to ensure that the agents which are used for this active treatment do not in themselves contribute to a caries problem. All too often patients are advised to drink fluids or suck boiled sweets or mints as a *pleasant* method of alleviating symptoms. However, agents which contain either sugar or acid should be avoided if at all possible in these subjects whose oral fluid clearance is low and in whom salivary buffering and remineralising is poor. It is no surprise that the areas most susceptible to root surface decay[27] are also those where sucrose levels are highest, and the clearance rate lowest, within the oral environment during chewing,[28] and where resting plaque pH is lowest.[29]

Quite apart from their xerostomic activity drug usage can also influence dental caries through the sugar content of the drugs themselves. The problems of sugars in medicines are well documented in children and are covered elsewhere in this text. However, there may also be some difficulty in adults who, for whatever reason, require medicines in a syrup form. In addition, there are drugs which contain relatively large quantities of sugar in their formulation. The most widely publicised example was that of the chewable antacid preparation *Gaviscon* which was associated with high caries activity.[30] Fortunately, the manufacturer of this product rapidly responded to the problem and the formulation has been changed to remove the sugar component. Unfortunately, there are still chewable antacids in the British National formulary that contain significant quantities of sugar (e.g. *Algicon*). In addition, there are a large number of over-the-counter (OTC) drugs which contain sugar, the extent of whose use is unknown, but whose sales are high.

It was demonstrated earlier that in the fit, healthy, unmedicated older individual, salivary flow is no different from that in the young. It is of interest however, that salivary glucose clearance is dramatically slower in older subjects than in the young. In addition glucose clearance in the hospitalized elderly is slower than in home-living subjects.[31] The authors of this report were able to correlate the rate of glucose clearance with the salivary secretion rate (despite there being no differences for secretion rates between the groups) and chewing frequency in addition to Plaque Index scores, and salivary counts for both Lactobacilli and *Strep. Mutans*.

Much of this section has dealt with the development of caries in older subjects, with particular emphasis on root surface lesions. Coronal caries is still a problem in the adult, indeed in the three-year study of elderly Iowan's quoted earlier[14] the annualized coronal caries increment was 1.2 surfaces per hundred surfaces at risk compared to 1.8 for root surfaces. In this same study 22 per cent of the sample did not develop any carious lesions and a further 13 per cent only developed root surface cavities. By contrast 34 per cent developed new coronal lesions and 31 per cent both coronal and root surface lesions. This evidence tended to confound results from cross-sectional epidemiological studies relating coronal and root surface lesions, posing the question of some variation in aetiology between these two disease processes.

Dental caries is an active disease in adults, and there is a suggestion that caries susceptible subjects remain so throughout their lives. Indeed Leske and Ripa[32] reported that 57.6% of their sample over the age of 55 developed one or more new root surface lesions over a three-year period. The vast majority of such lesions (82% of the new lesions) developed in subjects who already had one or more decayed or filled root surface. When new surfaces become exposed into the mouth through a combination of age and periodontal disease these too will be subject to attack.

There is little doubt that dietary sugars contribute to caries risk in adults, which may in turn be elevated in older subjects by a combination of age changes and disease. Continued dietary vigilance will be required along with specific prevention regimes, to ensure that caries activity is minimised in this second vulnerable element within society. An element which is going to grow in number, is going to undergo improved dental status and is likely to be more dentally aware and less willing to accept extraction and replacement with dentures in years to come.

Periodontal Disease

The single over-riding factor in the development of periodontal disease is the failure on the part of the subject concerned adequately to remove plaque from the surfaces of their teeth.

With one or two exceptions, there is little evidence that diet has a significant role in modifying the patterns of periodontal destruction. The concept that fibrous food can act as a natural toothbrush in man has been shown to be unfounded,[33,34] probably as a result of the bulbous shape of man's teeth which tends to shelter the gingival crevice from food during mastication.

It is not surprising that in some severe vitamin deficiency states (notably vitamin C which is intimately involved in collagen metabolism, possibly vitamin A which is involved in maintenance of mucosal integrity and Folic acid due to its action during DNA synthesis) periodontal integrity can be compromised. Indeed tooth mobility and loss is one of the early clinical signs of vitamin C deficiency. Fortunately, severe dietary deficiency of these vitamins is rare although local folate supplementation in the form of a mouthrinse can be of some benefit in pregnancy induced gingivitis where local folate deficiency has been postulated as an aetiological factor.[35,36] Scurvy (vitamin C deficiency) is rare, although it can occur in old people living on restricted diets.

There is a relationship between sucrose intake and quantitative measures of plaque deposition,[37,38] but there is little evidence that this has any effect on gingival inflammation and bony destruction. This effect is probably related to qualitative variations in the plaque, and that these measures of plaque relate to supra rather than sub gingival deposits, which may be of lesser significance in inducing the destruction of periodontal tissues during the disease.

An inverse relationship has also been demonstrated between fluoride in the water supply and periodontal destruction.[39] Whilst it is possible that this is a manifestation of altered bony metabolism in those who were life-long residents in a community with a fluoridated water supply, it is more likely that the reduced number of restorations in the teeth of those from a fluoridated community would reduce the frequency of artificial stagnation areas produced around poorly contoured restorations, especially in the areas between the teeth.

Diet has a limited role in the pathogenesis of periodontal disease. Periodontal problems only manifest themselves in the presence of severe dietary restriction.

Tooth Wear

The wear of the functional surfaces of teeth is a normal feature of ageing as a product of a life-times function in masticating food. However, there are some circumstances when the rate of loss of tooth tissue is accelerated as a result of dietary factors. Control of such factors usually leads to cessation of the abnormal wear.

Abrasive diet. Primitive methods of preparing food, particularly milling flour using sandstone wheels, or a stone mortar, results in a product which is highly abrasive in nature. Regular consumption of such a diet, or one in which large quantities of tough/fibrous food have to be chewed, will result in a characteristic pattern of toothwear. This produces broad, flat functional surfaces to the teeth, ideal for grinding food using a lateral sliding movement of the mandible (Fig 4). There is also a tendency for the dental arch to be shorter in such individuals as a result of wear in the contact areas between the teeth reducing the point contacts between two convex surfaces seen in young teeth to broad flat areas of contact after wear.

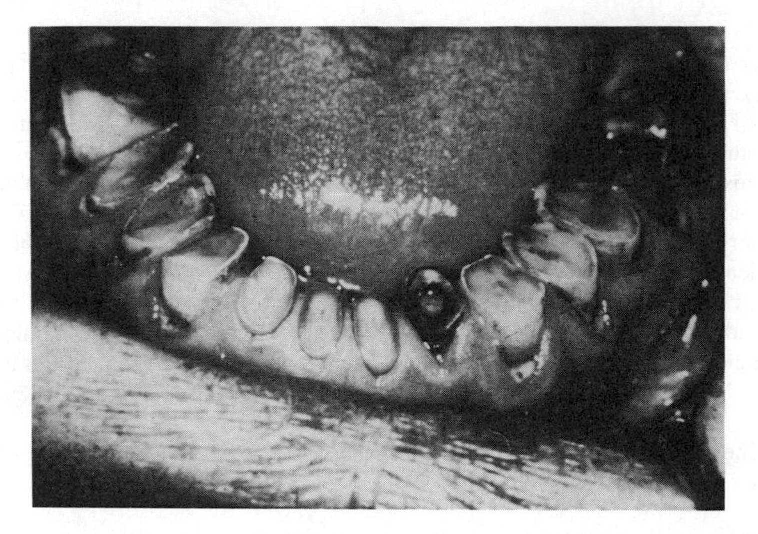

Figure 4. Broad, flat occlusal surfaces of teeth as a result of functional wear with an abrasive diet. The abrasive agent in this case was likely to be stone ground maize flour

It has been argued that this pattern of wear has a protective effect upon the teeth, primarily by removing the caries prone fissures and proximal surfaces of the teeth at an early stage in the mouth. The patterns of dental disease seen in populations with this type of diet are typified by some periodontal destruction and root caries lesions, but little coronal decay. It should be remembered that intake of refined carbohydrate is likely to be low in a population where methods of food preparation are so limited. It is important to note that root surface lesions can and do develop in people using a high starch, low carbohydrate diet.[17,18,40]

Table 2

Sources of acid from diet that have been implicated in producing advanced tooth wear.

Citrus fruits, Apples, Bananas
Fruit juices
Carbonated drinks (especially Cola and IrnBru)
Fruit flavoured infant drinks
Acid sweets (acid drops, mints, sour balls)
Vinegar, Pickles and Pickled onions
Liquid oral medicines (especially *Iron Tonics*)
Aspirin (especially soluble/chewable forms)
Vitamin C (soluble and chewable forms)

Erosive Diets. Teeth are fundamentally a complex calcium phosphate crystal, and as such will dissolve when in contact with acid. This is the basic mechanism behind the development of caries through acids produced as a result of the metabolism of fermentable carbohydrates in plaque. In addition, exposure of the surfaces of teeth to extrinsic acids from whatever source, and diet is only one, can result in bulk loss of tooth tissue. The sources of acid within a diet are numerous (Table 2). Obviously, it is impracticable to eliminate acid containing foods from a balanced diet, so once again it is a question of moderating dietary intake. It is well established that demineralised tooth surfaces can and do remineralise in contact with saliva. This pattern of repair will occur irrespective of the cause of the demineralisation, and if allowed to proceed will not result in tooth surface loss. However, should the frequency or the severity of the acidic challenge exceed the repair capabilities of the saliva, loss of tissue will occur.

The pattern of loss of tooth tissue depends upon the nature and location of the acidic challenge.

Relatively low levels of acid intake can contribute to accelerated wear in a subject who clenches or grinds their teeth, especially when the wear process has extended through the enamel, exposing dentine. Demineralised dentine is worn away at a faster rate than demineralised enamel producing an appearance akin to that of molar teeth in ruminants, where the enamel stands proud of the dentine giving a cup-like morphology to the tooth surface (Fig.5). The enamel margins are often jagged as they are unsupported and thence weak, and the surface of the dentine has a pearl-like texture indicative of erosive attack. When the secondary dentine is exposed, this can be raised slightly above the level of the primary dentine, a manifestation of the greater mineral component in this secondary dentine.

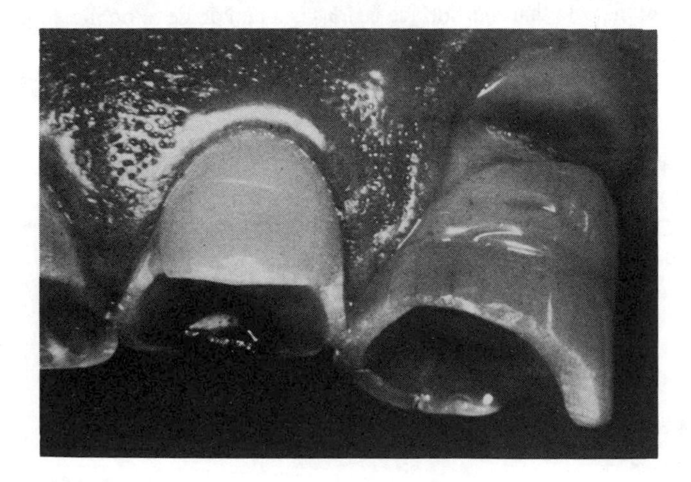

Figure 5. Cup-like loss of tooth tissue from the incisal edges of two upper anterior teeth. Note the amalgam restoration in the lateral incisor which is standing proud of the surrounding tooth tissue.

In contrast, frequent exposure to high levels of acid (e.g. citrus fruits/drinks, carbonated drinks, acidic drugs) can result in extrinsic bulk loss of tooth in the area exposed to the acid. Such exposure is commonly linked to a habit, thus High[41] reported extensive loss on the buccal aspects of teeth associated with a habit of retaining a carbonated drink in the buccal cheek pouch; where the ingestion of citrus fruits/drinks is a problem the maxillary anterior teeth are often involved.[42,43] Erosive lesions tend to be smooth with early loss of surface contour from the enamel. Once again the enamel may have a pearl-like appearance indicative of an acidic attack.

There are little data available about the prevalence of erosive wear in the population. One recent paper has estimated that 20 per cent of 26-30 year olds and 23 per cent of 46-50 year olds exhibited some evidence of erosion, although the proportion with severe erosion was greater in the older age-group (13 per cent as opposed to 8 per cent).[44] There is a perception amongst dental practitioners that the prevalence of toothwear is increasing, especially amongst the young.[45] It has been suggested that this apparent increase is linked to dietary change amongst children in association with increased consumption of carbonated drinks.

There are little data available concerning threshold levels for pH, or frequency of consumption, at which erosion commences. Stephan[46] demonstrated that the demineralisation of enamel commences at pH 5.5. Many soft drinks, especially fruit juices[47] and carbonated drinks[48] have pH values well below this level, as will citrus and other acid-containing fruits. There is some evidence to suggest that the demineralising potential of acids may vary from acid to acid, and in association with the pH level to which the acid is buffered.[49,50,51,52]

Järvinnen *et al*[53] reported a case control study assessing risk factors in cases of dental erosion. The adjusted odds ratios (the chance that a risk factor would be

Table 3

Adjusted odds ratios and population attributable risk for dietary factors in subjects with clincally apparent erosive tooth wear[53]

Factor	Adjusted Odds Ratio	Population attributable Risk (%)
Citrus fruits (>x2 daily)	37	26
Apple vinegar (weekly or greater)	10	15
Soft drinks (4-6 times or more per week)	4	26
Sports drinks (weekly or greater)	4	15

recorded in a subject with clinical signs of erosion) and the adjusted population risks (the percentage by which clinically apparent erosion would decrease if the risk factor were removed) for dietary factors are given in Table 3. Consumption of citrus fruits was by far the most common risk factor although the population attributable risk for soft and sports drinks combined was 41 per cent. When the same authors assessed the odds ratios for erosive wear compared to the frequency of intake of citrus fruits and/or soft drinks, the ratios rose dramatically when the frequency of consumption rose to 2 or more times daily. Interestingly, these authors also reported a relationship between erosive wear and low resting salivary flow (<0.1mL/min) which support work reported elsewhere.[54,55]

4 SUMMARY

Diet can and does play an important role in influencing the pattern of dental disease in adults. Dental caries continues to be a problem in adults. Caries risk is likely to escalate with increasing age as a result of age-associated changes in salivary clearance of sugars, an increase in the numbers of surfaces at risk through root exposure and alterations in diet with a tendency to consume more non-milk extrinsic sugar. Superimposed upon these dietary/host factors is the possibility of alteration in salivary flow and composition as a result of disease and/or drug usage in addition to changes in composition as a product of ageing alone. Dietary counselling and preventative care should be an integral part of dental management of adults as well as children, especially the old. Saliva also has a central role in the protection of teeth from acid erosion. Tooth wear can present a complex management problem in the adult. Appropriate dietary advice should be given to the young and old in an endeavour to reduce the frequency of intake of dietary acids, especially during early adult life, to maintain adequate quantities of tooth tissue to provide function for the life-time ahead.

In the absence of malnutrition diet has a minimal role in the aetiology of periodontal disease.

Control of dietary excess, in terms of frequency of intake rather than necessarily absolute quantity remains a cornerstone for maintenance of dental health.

5 REFERENCES

1. J.E. Todd and D. Lader, Adult Dental Health 1988: United Kingdom, HMSO, London, 1991, 9.
2. OPCS Social Trends 18, HMSO London 1988
3. G. Agerberg and G.E. Carlsson, Acta.Odontol.Scand. 1981, 39, 147.
4. H. Kaplan, Geriatrics, 1971, 26, 96.
5. O. Ramsey, J.Prosthet.Dent. 1983, 49, 16.
6. H. Wayler and H.M. Chauncey, J.Prosthet.Dent. 1983, 49, 427.
7. H-S.J. Gunne, Acta Odontol.Scand. 1985, 43, 139.
8. W.T.C. Berry, Dent Pract.1972, 22, 249.
9. B.E. Gustafsson, C.E. Quensel, L.S. Lanhe, C. Lundquist, H.Graham, B.E. Bonnow, B.Krasse. Acta.Odontol.Scand. 1954, 11, 232.
10. A. Scheinin and K.K. Mäkinen. Acta.Odontol.Scand. 1975, 33, 1.
11. T.M.Marthaler and E.R. Froesch, Br.Dent J, 1967, 123, 597.
12 E. Newbrun, C. Hoover, G. Methaux, H. Graff. J Am Dent Assoc. 1980, 101, 619.
13. F.J. Fisher, Brit Dent J. 1968, 125, 447.
14. J.S. Hand, R.J. Hunt, J.D. Beck, Gerodontics, 1988, 4, 136.
15. A.S. Papas, C.A. Palmer, R.B. McGandy, S.C. Hartz, R.M. Russell. Gerodontics, 1987, 3, 30.
16. M.M. Vehkalahti, I.K. Paunio. J Dent Res. 1988, 67, 911.
17. R.G. Schamschula, P.H. Keyes, R.W. Hornabrook. J Am Dent Assoc, 1972, 85, 603.
18. R.G. Schamschula, D.E. Barnes, P.H. Keyes, W. Gulbinat. Community Dent Oral Epidemiol, 1974, 2, 295.
19. A.J. Rugg-Gunn, M.A. Lennon, J.G. Brown. Br Dent J.1986 161, 359.
20. Dept. of Health. Dietary sugars and human disease. Report on health and social subjects HMSO London, 1989.
21. D.J. Witter, A.B. Cramwinckel, G.M.J. van Rossum, A.F. Käyser. J.Dent. 1990, 18, 185.
22. P.C. Fox, K.A. Busch, B.J. Baum. J Am Dent Ass, 1987, 115, 581.
23. J. Scott, Gerodontics, 1986, 5, 149.
24. B.J. Baum, J.A. Ship, A.J. Wu. Crit Rev Oral Biol Med, 1992, 4, 53.
25. M.U. Rashid and D.N. Bateman. Brit J Clin Pharmac, 1990, 30, 25.
26. S. M. Levy, K.A. Baker, T.P. Semla, F.J. Kohout. Gerodontics, 1988, 4, 119.
27. R.V. Katz, S.P. Hazen, N.W. Chilton, R.D. Mumma. Caries Res, 1982, 16, 265.
28. C. Dawes and L.M. MacPherson. J Dent Res, 1993, 72, 852.
29. I. Kleinberg and G.N. Jenkins. Arch Oral Biol, 1964, 9, 493.
30. H.D. Glenwright, L. Shaw, C. Cooke. Brit Dent J, 1988, 164, 6.
31. J.C. Hase, D. Birkhed, M-L, Grenment, B. Steen. Gerodontics, 1987, 3, 146.
32. G.S. Leske, L.W. Ripa J Pub Hlth Dent, 1989, 49, 142.
33. J.A. Reece and J.N. Swallow. Brit Dent J, 1970, 128, 535.
34. P. Longhurst and D.S. Berman. Brit Dent J, 1973, 134, 475.
35. A.R.C. Pack and M.E. Thomson. J Clin Periodontol, 1980, 7, 402.
36. M.E. Thomson and A.R.C. Pack. J Clin Periodontol, 1982, 9, 275.
37. K. Paunio, K. Markinen, A. Scheinin, K. Ylitalo. Acta Odontol.Scand, 1975, 33, 217.

38. E.M. Rateitschak-Pluss and B. Guggenheim J Clin Periodontol, 1982, 9, 239.
39. H.R. Englander, R.G. Kesel, O.P. Gupta. Amer J Pub Hlth, 1963, 53, 1233.
40. J.L. Hardwick. Br Dent J 1960, 108, 9
41. A. High. Br Dent J, 1977, 143, 403.
42. K.J. Lewis and B.G.N. Smith. Br Dent J, 1973, 135, 400.
43. M.E. Reuter. Br Dent J, 1978, 145, 274.
44. A Lussi, M. Schaffner, P. Hotz, P Suter. Community Dent Oral Epidemiol, 1991, 19, 286.
45. R.S. Levine. J Dent, 1973, 2, 85.
46. R.M. Stephan. J Am Dent Assoc, 1940, 27, 218.
47. L.Z.G. Touyz and R.M. Glassman. J Dent Assoc S Afr. 1981, 36, 195.
48. T.D. Eccles and W.G. Jenkins. J Dent, 1974, 2, 153.
49. F.J. McClure and S.J. Ruzicka. J Dent Res, 1946, 25, 1.
50. W.N. Elsbury. Br Dent J, 1952, 93, 177.
51. C.J. Kleber, M.S. Putt, C.J. Muller. J Dent Res, 1979, 58, 1564.
52. J. Meurman, I. Rytömaa, K. Kari, T.T. Laokso, H. Murtömaa. Caries Res, 1987, 21, 353.
53. V.K. Järvinnen, I.I. Rytömaa, O.P. Heinönen. J Dent Res, 1991, 70, 942.
54. I. Hellström. Scand J Dent Res, 1977, 85, 71.
55. J.M.M. Wöltgens, P. Vingerling, J.M.A. de Blieck-Hoger-Vorst, D.J. Berovets. Clin Prev Dent, 1985, 7, 8.
56. Oral Health for All 2000, 1986 World Health Organisation
57. R.V. Katz. J Dent Res, 1985, 63, 814.
58. J.G. Steele, Personal Communication 1993.
59. M.A. Donachie and A.W.G. Walls. J Dent Res, 1991, 70, 684.
60. S.P. Hazen, N.W. Chilton, R.D. Mumma. J Dent Res. 1972, 51 (Suppl), 219.
61. W.G. Lohse, H.G. Carter, J.A. Brunelle. Millit Med. 1977, 141 700.
62. A.J. Miller et al. Oral Health of United States Adults. NIH No 87, 1987, 28
63. J.W. Stamm and D.W. Banting. J Dent Res, 1985, 59 (Special Issue) 405.
64. M.M. Vehkalahti, M. Rajala, R. Tuomien, I. Paunio. Comm Dent Oral Epidemiol. 1983, 11, 188-190

A Dietitian's View of a Sugarless Diet

M. Sanderson

DIVISION OF APPLIED CHEMISTRY, LIFE SCIENCES AND POLYMER TECHNOLOGY, UNIVERSITY OF NORTH LONDON, 166–220 HOLLOWAY ROAD, LONDON N7 8DB, UK

1 THE TASK OF A DIETITIAN

In the area of public health, the role of the dietitian is not only to engage in the debate but to also translate dietary recommendations into practical guidelines for the general public. Translating sugar guidelines has presented dietitians with a great deal of difficulty for a number of years. The official UK recommendations on sugar consumption have been extremely vague. They have consisted of messages to consume less or cut down sugar intake. For many people this was a less than helpful message: if they were taking very little sugar, should they still cut down? For those taking large quantities of sugar, what sort of reduction was necessary to achieve the desired effect?

2 NUTRITIONAL GUIDELINES

The NACNE report[1] was the first British report to give quantified guidelines for sugar, but the recommendation caused so much controversy that they were never officially adopted, although they did become incorporated in many health authority documents. Now, 10 years later, we at last have clear and unequivocal quantative guidelines on sugar consumption. These were published two years ago by the Government's COMA Panel in its report on Dietary Reference Values[2] . Not only did they give guidelines on the quantity of sugar to be consumed but also clearly defined the different types of sugars in the diet - non-milk extrinsic sugars (NME), milk and intrinsic sugars.

Non-milk extrinsic sugars are sugars that have been extracted from either root, fruit or stem of a plant. Previously, these sugars were often rather loosely called "added sugars". Intrinsic sugars are those which have not been removed and are eaten as an intergral part of the whole fruit or vegetable. Therefore, the sugar in the juice squeezed from an orange is extrinsic but if eaten as part of the whole fruit, intrinsic. Fruit juice added to foods as a sweetener would, under this classification, be defined as extrinsic sugar.

The COMA panel[2] recommended that the consumption of NME sugars should contribute not more than 10% of total energy or 11% of food energy. It was also stipulated that whatever the energy intake of the individual, the total daily intake should not exceed about 60g of NME sugars. This quantity is the equivalent of 12 tsps of this type of sugar, derived from both manufactured foods and that added to foods and drink by the consumer.

These recommendations have been widely accepted by health professions and most of the food industry. But, there are those who are still dispute these recommendations. In support of their argument, they quote the most frequently cited sentence in the Dietary Reference Values report[2] - "data in support of any specific quantified targets for non-milk extrinsic sugars were scanty".

This is both untrue and wrong. There is enormous evidence on quantified guidelines. The epidemiology of sugar consumption and dental caries is one of the most assiduously studied areas all over the world. Professor Aubrey Sheiham has already published two reviews of the evidence in this area, once in 1983[3] and again in 1991[4]. He concluded that on the basis of the epidemiological evidence, to reduce the incidence of dental caries, more rigorous quantative guidelines than those suggested by the COMA panel[2] were necessary (even with optimum fluoridation).

However, other scientists have also examined the data. Cannon and colleagues[5] analysed a hundred published reports giving nutritional guidelines. Of these 100, 82 recommended a reduction in sugar consumption and 23 also gave quantative guidelines; the average figure quoted for quantative guidelines was that sugar should provide no more than 10% of energy. In addition, the last two reports from the World Health Organization for Europe[6] and the World[7], have made recommendations much on the same lines as those in Britain. There does, therefore, appear to be a consensus on the level of intake compatible with health.

3. ACHIEVING TARGETS FOR SUGAR CONSUMPTION

If we accept this target of 10% of energy what changes are needed in the diet?

Estimating the consumption of NME sugars in the UK is beset with many difficulties. COMA, in its report on Dietary Sugars and Human Disease[8] in 1989, estimated the consumption of total sugars to be between 18% to 20% of energy. Later reports by COMA[2] have put the consumption of NME sugars at between 13% to 15% of energy and the Health Education Authority[9], in their literature, at between 15 to 20% of total energy. Many believe these latter estimates are probably conservative as they are calculated from the amount of NME sugars available for consumption, minus a wastage factor. The wastage factor for sugar has been variously estimated at between 30 and 50%.

If the more conservative consumption level estimates of the COMA panel[2] and the Health Education Authority[9], are taken, to meet the DRV's on NME sugars will require the population of Britain to reduce its current consumption by 20-30%. This is not an insignificant amount. In order to achieve this reduction, any health education messages and structural changes must be tailored to the diets of the British public.

Of all the sugar available for consumption in the UK, nearly 74% is used by industry, predominately the food industry; only 26% goes directly to the consumer via packet sugar. Packet sugar has often been referred to as "visible sugar" since this is the sugar consumers add to food themselves.

Table 1 Sucrose consumption (kg/head/year) in the U.K. between 1941 and 1991, and the percentage of this sold as "visible" packet sugar.[8]

Year	Sucrose Consumption (Consumption Level Enqiry)	% "visible"
1941	30	-
1946	34.8	-
1951	42.0	-
1956	48.6	54.5
1961	50.1	53.3
1966	47.9	52.4
1971	45.5	50.4
1976	40.8	44.0
1981	37.8	43.9
1986	37.3	29.7
1991	36.0	26.2

The relationship between NME sugars and visible sugars is interesting. Since 1956 there has been a steady decline in the amount of NME sugars going direct to the consumer (Table 1). From contributing about half of the NME sugars consumed, it now contributes just over a quarter. In terms of health education, advising the public to cut down on the amount of sugar they add to food will now have only a limited effect.

Examining the distribution of sugar to industry will give some indication of the contribution to NME sugars made by the various manufactured products in the diet (Table 2).

Table 2 Where Sugar goes in the Food Industry
Data from U.K. Sugar Industry Statistics 1991

	%
Soft drinks	25.4
Confectionery	25.2
Baker's, Biscuits & Cereal products	19.7
Convenience foods	9.8
Preserves	6.0
Icecream & Milk products	4.6
Beers & Wine	3.4
Pharmaceuticals	1.9
Other	4.0
	100

The contribution from the pharmaceutical industry appears quite small. Icecream, milk products and preserves are not at present major contributors of sugar to the British diet. However, soft drinks provide just over 25%, confectionery about 25%, and baked, biscuit and cereal products almost 20%. These three sectors account for over 70% of all industrial sugar and so provide the biggest contribution of sugar to the British diet. One way, therefore, of achieving the DRV guidelines would be to advise consumers to reduce the quantity of sugar consumed from these three commodities. If consumers were to reduce their intake of sugar from these three groups by 50%, this would lead to an overall reduction in total NME sugar intake of approximately 25%. As the decrease in NME sugars needed to meet the DRV recommendations is between 20-30%, this is one strategy that could be used to realize this target. This reduction can be achieved either by altering the NME sugar content of these products by 50% or by reducing the consumption of these products by 50%.

In some sectors, reducing the sugar consumed in products by 50% might not be a difficult problem - there is now a wide range of low sugar drinks on the market and the NME sugar content of fruit juice can be reduced per unit volume by diluting it with water. There are also many cereals and bakery products with a low sugar content, and a widening range of snack foods. There are, at present, very few lower sugar confectionery products available for the consumer in the UK, so here a substitution of product may have to be made.

Therefore, to meet government recommendations, there is a need to reduce NME sugar consumption by 20-30%. This could be acheived in a number of ways but, in terms of health education messages, one of the simplest is to encourage a reduction by 50% of the quantity of sugar consumed through the 3 highest sugar-containing groups of foods - soft drinks, confectionery and baked and cereal products.

4 IMPLICATIONS OF REDUCING SUGAR CONSUMPTION

If the DRV reccommendations on sugar are met, and if the diet is altered in no other way, this will result in a reduction in energy intake, of varying degrees, for the majority of the population. For some, it may be necessary to replace this energy, but for others it may not be necessary.

Adamson[10] has shown that children in the North East obtain between 17 to 18% of their energy from NME sugars. If these children met the recommendations, this would represent a daily reduction in energy intake of between 123 and 138 kcal per day (Table 3). Children need energy for growth and development, and so these kcals need to be replaced.

Table 3 The energy and sugars intake of 12-year-old children[10], compared with the dietary reference values given by the Department of Health[2]

	Boys	Girls
Total energy intake, kcals	2060	1974
Energy from NME sugars, % energy	17	18
Energy from NME sugars, kcal	350	355
DRV, % energy	11	11
DRV, kcal	227	217
Difference, kcal	123	138

The percentage of energy that adults derive from NME sugars has been estimated to be approximately 14%[9]. Conforming to the recommendations for this group would mean an average reduction of between 67 and 98kcals per day (Table 4).

Table 4 The energy and sugars intake of adults[11], compared with the dietary reference values given by the Department of Health[2]

	Men	Women
Total energy intake, kcal	2450	1680
Energy from NME sugars, % energy	14	14
Energy from NME sugars, kcal	345	235
DRV, % energy	10	10
DRV, kcal	245	168
Difference, kcal	98	67

For approximately 50% of the population, this small reduction in kcals could make an important contribution to their health. Latest estimates suggest that the percentage of the population who are overweight or obese is still increasing. The most recent results suggest that in England, 51% of men and 41% of women are now overweight, a worrying statistic in terms of public health (Table 5).

Table 5 Prevalence of overweight

Year	Men	Women
1981[12]	39%	32%
1990[11]	45%	36%
1993[13]	51%	41%

For the proportion of the population who are not overweight or obese, the lost kcals will need to be replaced.

In order to conform to other Government nutritional guidelines[2], the replacement kcals for both children and adults should be derived from an increased intake of complex carbohydrates and intrinsic sugars. For example, the quantities of bread, cereals, pasta and pulses should be increased as well as fruit and vegetables.

For those who need it, the replacement of the energy deficit is not difficult. In practical terms, the energy deficit in adults can be replaced by a slice of bread and a piece of fruit. Children can satisfy their energy needs by extra bread and potatoes or larger helpings of rice and potatoes (Table 6).

Table 6 Typical energy values[14]

		kcals
Fruit -	apple/orange/pear	50
	banana	100
1 small slice bread		50
2 crispbread		50
1 small potato		50
Good helping spagetti		200
1 helping boiled rice		100

In the past, advice on how to increase the energy of the diet, while keeping to other guidelines, has not always been given. The result has been that, while consumers have managed to keep their sugar intake down, in some cases the fat content of the diet has increased: a phenomenon often referred to as the "sugar-fat" seesaw.

The point is often made that people rarely just eat dry bread, it needs moistening or lubricating. The usual agent for this is fat. Similarily plain boiled potatoes are not the most attractive of options, chips or crisps are far more appealing. But this can be countered by giving advice to eat more bread by eating thicker slices, to eat bigger and fatter chips to cut down the potential surface area available to fat, and to eat the lower fat varieties of products such as potato crisps.

The more enlightened food manufacturers have already taken this message on board and are producing products that are both lower in fat and sugar. Examples of this can be seen in the biscuit sector where one manufacturer has produced a digestive biscuit that is both lower in fat and lower in sugar than the standard product. Similarily, in the dairy and frozen dessert sector, products now coming onto the market are lower in sugar and fat. There are still, unfortunately, some less technically advanced manufacturers who, in responding to the Health of the Nation initiative, have produced lower-fat products but, at the same time, have managed to produce a product that is higher in sugar than the standard one. However, with good nutritional advice and information on how to interpret the existing nutritional labelling, the consumer can avoid this particular pitfall.

In the future, the "sugar-fat" seesaw need not be an inevitable consequence of consumer attempts to reduce sugar consumption. Indeed, the Nutrition Task Force, set up to look at ways and means of implementing the Health of the Nation initiative, are very aware of the potential nutritional consequences of food manufacturers concentrating on only one nutritionally-related disease and are recommending that any product re-formulation should take into account all other COMA recommendations.

5 ROLE OF ALTERNATIVE SWEETENERS

Finally, a question that is often asked of dietitians is - what are the health consequences of increasing the uptake of alternative sweeteners? Are we substituting one set of health problems for another? Due to their gastrointestinal effects, the intake of bulk sweeteners is probably self-limiting. Many new products containing these types of sweeteners now clearly indicate the maximum daily intake of the product. Apart from their unpleasant effects on the gut, these bulk sweeteners appear to have no other health hazards.

There has been, by contrast, much debate over the health problems associated with intense sweeteners. There is, however, no evidence to indicate that any of the sweeteners permitted in the UK are carcinogenic in humans nor that they cause any other serious health problems in a healthy population.

But the use of artifical sweeteners, particularly the intense sweeteners does maintain a high threshold for sweetness. There are many that would argue that, in the long-term interests of health, this threshold should be reduced. They would advise that alternative sweeteners should only be be used as a bridging measure to allow consumers to gradually adapt to a lower threshold for sweetness.

6. CONCLUSION

To conclude, from the dietitian's point of view, the publication by COMA[2] of clear quantative guidelines on sugars consumption has enabled us to give much more relevant and unambiguous advice. The recent trends in manufacturing practice are enabling consumers to put this advice into practice without increasing other unfavoured nutrients. The prospects of implementing a sugar health policy at present look more hopeful than at any other time. We are probably not yet at the stage of achieving a sugarless diet and, indeed, this might not be desirable, but a less sugared diet is now achievable without compromising other aspects of the diet.

REFERENCES

1. Health Education Council, 'Proposals for nutritional guidelines for health education in Britain', NACNE, London, 1983.

2. Department of Health, 'Dietary reference values for food energy and nutrients for the United Kingdom', Report on health and social subjects 41, HMSO, London, 1991.

3. A. Sheiham, Lancet, 1983, 1, 282.

4. A. Sheiham, Br. Dent. J., 1991, 171, 63.

5. M.D.C.M. Freire, G. Cannon, A. Sheiham, 'Sugar and health' Monograph Series 1992, No.1, University College, London.

6. W.P.T. James, 'Healthy nutrition' W.H.O. Regional Publication European Series No.24, 1988.

7. W.H.O., 'Diet, nutrition and the prevention of chronic diseases', W.H.O. Technical Report Series 797, 1990.

8. Department of Health , 'Dietary sugars and human disease', Report on health and social subjects 37. H.M.S.O., London, 1989.

9. Health Education Authority, 'Sugars in the diet', London, 1990.

10. Report of an expert working group, 'Nutrition guidelines for school meals'. The Caroline Walker Trust, 1992.

11. J.Gregory, K. Foster, H. Taylor, M. Wiseman, 'The dietary and nutritional survey of British adults', H.M.S.O., London, 1990.

12. OPCS 'Heights and Weights of Adults in the U.K.' H.M.S.O., London, 1981.

13. A. White, G. Nicolaas, K. Foster, F. Browne, S. Carey, 'Heath Survey for England 1991' OPCS. H.M.S.O., London, 1993.

14. A, Micklewright, V. Todorovic, 'A pocket guide to clinical nutrition'. PEN Group of the British Dietetic Association, 1989.

Helping the Consumer to Make Better Choices of Medicines

I. C. Mackie

DEPARTMENT OF ORAL HEALTH AND DEVELOPMENT, UNIVERSITY DENTAL HOSPITAL, HIGHER CAMBRIDGE STREET, MANCHESTER M15 6FH, UK

1. INTRODUCTION

Dental caries causes considerable pain and distress especially to young children, and its prevalence is related to the intake of refined carbohydrate. It is surprising to record how many medicines, supposedly produced to help young children, may exacerbate dental caries because of their sugar content. Hobson[1] highlighted the problem of sugar-containing medicines being a contributory factor to dental caries, especially in chronically sick children who require long term medication. Maguire[2] reinforced the importance of this problem when she reported that many preparations prescribed for long-term usage were sugar-based. She recommended that manufacturers of liquid generic preparations should be encouraged to provide suitable sugar-free alternatives. Mitchell[3] made three suggestions to reduce the use of sugar-containing medicines. First that all sugar-containing medicines should be labelled with the concentration of sugar specified in g/ml. Second, pharmaceutical manufacturers should whenever possible replace sucrose in their formulations, and third, all medical practitioners should in the first instance prescribe sugar-free preparations rather than those containing sugar.

Suggestions such as these have usually been made in relation to prescribed medicines, however over-the-counter medicines seem to have been ignored.

2. OVER-THE-COUNTER MEDICINES

When the type of medicines people use on a regular basis are investigated, the findings are most intriguing. Dunnel and Cartwright[4] found that 74 per cent of children below the age of 2 years had received a non prescription medicine in the 2 weeks prior to their investigation. Fry, Brooks and McColl[5] reported that on any one day probably 60 per cent of the population took some type of medication with 27 per cent being prescribed and the remaining 33 per cent being over-the-counter preparations. Brooks[6] stated that on any one day 17 per cent of children were likely to be taking an over-the-counter cough medicine. A more recent investigation by Rylance et al[7] found that on average children took medicines one week in every eight. Prescribed medicines accounted for 55 per cent of the preparations used, with the remaining 45 per cent being over-the-counter.

From these figures, it would appear that many children take medicines on a regular basis of which about a half are over-the-counter preparations. The most commonly used over-the-counter children's preparations include analgesics and cough medicines[5,7].

When the way these preparations are used is understood it is relatively straightforward to appreciate how they may contribute to the development of dental caries.

Children may be given a sugar based analgesic last thing at night to relieve pain so as to help them 'go off to sleep', or a night time 'tickly cough' may be eased by a sugary cough syrup. Sugary medicines will be given during the night when they wake up because of pain or a cough. Sugar given just before going to bed or during the night can be especially harmful to the teeth, as the natural cleansing action of saliva is limited because salivary secretion is reduced during sleep. Sugar therefore stays on the teeth for long periods, the acids produced are not buffered or removed, causing dental caries.

The pharmaceutical industry has become aware of the problem to dental health of medicines containing sugar and is producing an increasing number of sugar-free medicines for children. There were 106 sugar-free medicines listed by Brandon and Sadler[8] in 1985; this figure had risen to 156 in 1987[9] and 218 in 1988[10]. This is a welcome move; but are these sugar-free medicines being recommended by pharmacists, requested by customers and prescribed by doctors?

3. USE OF PAEDIATRIC MEDICINES

An investigation into paediatric sugar-containing and sugar-free medicines stocked and recommended by pharmacists in the North Western Region of England was undertaken over a one year period in 1990 - 1991[11]. A questionnaire was sent to each of the 769 pharmaceutical retail outlets in the North Western Region of England. The questionnaire contained a list of 34 named over-the-counter paediatric preparations, 14 contained sugar and 20 similar but sugar-free alternatives. The pharmacists were asked which preparations they stocked, which were best sellers and which they recommended for 12 specified childhood ailments.

The completed questionnaire was returned by 555 of the 769 pharmaceutical retail outlets, a response rate of 72 per cent. Of the 34 preparations named in the questionnaire, 14 were stocked by more than 90 per cent of pharmacists and of these only 2 were sugar-free. At the other end of the scale, 13 preparations were stocked by less than half of all the pharmacists, and all were sugar-free. Of the 7 preparations considered by the pharmacists to be best sellers only one was sugar-free. For the 12 childhood ailments it was found that for 7 of these, sugar-containing medicines were recommended statistically significantly more often than those which were sugar-free. Thus pharmacists in the North Western Region of England stocked and recommended more sugar-containing over-the-counter paediatric medicines than those which were sugar-free.

A recent project looked at general medical practitioners' prescribing habits in 4 districts in the North Western Region of England[12]. Prescription data were collected for 6 medicines which were commonly prescribed for children and which had readily available sugar-free alternatives. These were codeine phosphate, pholcodine, pseudoephedrine, paracetamol, amoxycillin and co-trimoxazole. The four districts were Oldham, Rochdale, Bury and Bolton. It can be seen from Table 1 that sugar-containing medicines were prescribed more frequently than their sugar-free alternatives. In Bury the difference was 12 per cent in favour of sugar-containing medicines but this rose to 28 per cent for Bolton.

The most commonly prescribed children's medicine was paracetamol and when this was considered, the differences between the percentages of sugar-containing and sugar-free were found to be even greater ranging from 26 per cent in Bury to 60 per cent in Bolton (Table 2).

TABLE 1: PERCENTAGES OF SUGAR-CONTAINING (SC) AND SUGAR-FREE (SF) PRESCRIPTIONS FOR THE 6 MEDICINES FOR APRIL-JUNE 1991

	Oldham	Rochdale	Bury	Bolton
SC	58	60	56	64
SF	42	40	44	36

TABLE 2: PERCENTAGES OF SUGAR-CONTAINING AND SUGAR-FREE PRESCRIPTIONS OF PARACETAMOL FOR APRIL-JUNE 1991

	Oldham	Rochdale	Bury	Bolton
SC	73	67	63	80
SF	27	33	37	20

Thus the doctors in the 4 districts prescribed a far greater proportion of sugar-containing paediatric medicines than the sugar-free alternatives.

In an investigation into general medical practitioners' views on prescribing sugar-free medicines for children it was found that their main consideration was whether the drug was effective, safe and acceptable to the child. When one was found to be effective they continued to use it. Most prescription writing was therefore described as automatic. Once a named drug was prescribed it continued to be so because it required positive thought to write something different. One thing that the doctors were worried about was compliance, they would not prescribe a sugar-free medicine if they thought that a child would not take the medicine. Yet, none had received negative comments about sugar-free medicines[13].

For dental health reasons, pharmacists should be encouraged to recommend sugar-free, doctors to prescribe sugar-free and customers or patients to request sugar-free. Currently these health professionals automatically recommend or prescribe preparations which usually contain sugar. Parents have reported that if given a choice between a medicine containing sugar and one without they would choose the sugar-free. Also they would accept a sugar-free medicine as an alternative to a sugar-containing medicine which they had requested if this was recommended by their pharmacist[14]. Thus there is a need for pharmacists and doctors to realise that consumers are interested in sugar-free preparations and should be prepared to recommend or prescribe them.

4. ENCOURAGING THE USE OF SUGAR-FREE MEDICINES

If Mitchell's[3] three suggestions are reconsidered, the first was that all sugar-containing medicines should be labelled with the concentration of sugar specified in g/ml. This is a relatively straight forward process, if the ingredients were given on the packaging, the consumer could check whether the medicine contained sugar. However, even if the manufacturers did agree to this, the consumer might still become confused if sweetening agents such as sucrose, fructose, honey, glucose, hydrogenated glucose syrup and sorbitol were listed. Many health professionals have problems deciding which sweetening agents are cariogenic and which are safe for teeth. Therefore, how are parents going to cope? A tooth-friendly logo might well be the answer to the problem of defining the cariogenicity of medicines. If the logo was stamped on the front of the packaging, consumers would know whether the medicine was safe for teeth, rather than having to try to decipher a complicated list of ingredients.

His second suggestion was that the pharmaceutical manufacturers should be encouraged to replace sucrose in their formulations. This would be ideal, but this recommendation was made as long ago as 1978 and many paediatric medicines still contain sugar. The pharmaceutical industry responds to consumer pressure[15]. The industry does not produce what concerned professional groups recommend, but instead it responds to what the consumer will buy.

The third suggestion was that medical practitioners should be encouraged to prescribe sugar-free preparations rather than those containing sugar. This suggestion ought to be broadened by including all health professionals: doctors, pharmacists, dentists and health visitors should be either prescribing or recommending sugar-free medicines.

5. HELPING THE CONSUMER TO CHOOSE SUGAR-FREE MEDICINES

A pilot dental health education campaign has been mounted in the North Western Region of England to encourage greater use of sugar-free paediatric medicines. The campaign sought to increase the proportion of sugar-free medicines prescribed for children by general medical practitioners, and recommended by pharmacists.

The campaign was conducted in 2 Family Health Service Authority areas with two others acting as controls in order to evaluate the campaign. The most important factor in deciding which areas to use was the availability of prescribing data, referred to as PACT data; Prescribing, Analyses and Cost. The PACT data were used to evaluate the campaign. Four Family Health Service Authority Areas were selected which had similar socio-demographic profiles and for which PACT data were available for the same time periods, so as to exclude seasonal variation. The four areas chosen were Oldham, Bolton, Bury and Rochdale. These were then matched in pairs, Oldham with Rochdale and Bury with Bolton. One from each pair was allocated randomly as the test area. The two test areas were Oldham and Bury.

The target groups for the campaign were general medical practitioners, pharmacists and their counter staff, health visitors, dentists and young mothers with children below 2 years of age. Dental health education material was specifically designed for each target group. The campaign slogan was "Smile for sugar-free medicines". The material consisted of leaflets for each group, inserts within the leaflets listing some sugar-containing medicines and the sugar-free alternatives, wall charts giving the same information, stickers for health professionals and an A3 poster for waiting rooms, shops, health educational displays and baby clinics. A pack of materials was produced for each doctor, pharmacist, health visitor, and dentist within the test areas.

Rather than send this material through the mail postgraduate meetings were organised for each professional group in the test areas. Each meeting consisted of a free buffet followed by a presentation entitled "A spoonful of sugar-free helps tooth decay go down". Sponsorship was obtained for each meeting from either Cupal, Reckitt and Coleman or Wellcome. All the meetings took place in February 1992. For those health professionals who did not attend a meeting an attempt was made to contact each in person and give them their "Smile for sugar-free medicine" pack together with a verbal explanation about the campaign.

For the numbers of prescriptions, data were collected for the six medicines which are commonly prescribed for young children and which have sugared and sugar-free alternatives. Baseline data were obtained for April to June 1991 for the four areas and post campaign data were for April to June 1992.

From Table 3 it can be seen that Oldham showed an increase in prescriptions for sugar-free medicines of 7 per cent from 42 in April-June 1991 to 49 per cent post campaign in April-June 1992. In comparison, Rochdale showed a 7 per cent decrease in sugar-free prescriptions from 40 to 33 per cent. For Bury and Bolton there were very little differences. Table 4 shows the most commonly prescribed paediatric medicine, paracetamol.

TABLE 3: PERCENTAGE OF SUGAR-FREE PRESCRIPTIONS FOR THE 6 MEDICINES FOR APRIL-JUNE 1991 AND 1992.

	Oldham (T)	Rochdale	Bury (T)	Bolton
April-June 1991	42	40	44	36
April-June 1992	49	33	43	37

TABLE 4: PERCENTAGES OF SUGAR-FREE PRESCRIPTIONS FOR PARACETAMOL FOR APRIL-JUNE 1991 AND 1992

	Oldham (T)	Rochdale	Bury (T)	Bolton
April-June 1991	27	33	37	20
April-June 1992	40	36	43	29

In the test area of Oldham sugar-free prescriptions for paracetamol increased by 13 per cent from 27 to 40 compared with an increase of 3 per cent in Rochdale from 33 to 36. However, in the test area of Bury the increase of 6 per cent from 37 to 43 was less than that for Bolton where sugar-free prescriptions increased from 20 to 29 per cent.

The results from this pilot campaign are somewhat equivocal. It was encouraging that in one test area, Oldham, there was a 7 per cent increase in the prescription of sugar-free medicines.

6. CONCLUSION

The dental profession and parents would prefer children to have sugar-free medicines instead of ones which contain sugar but in reality it is the sugar-containing medicines which are manufactured, prescribed, recommended and sold.

REFERENCES

1. P. Hobson, in "Sugarless - The Way Forward", ed. A.J. Rugg-Gunn, Elsevier Applied Science, London, 1991, 125-133.

2. A. Maguire, in "Sugarless - The Way Forward", ed. A.J. Rugg-Gunn, Elsevier Applied Science, London, 1991, 134-153.

3. G.M. Mitchell, in "Sugarless - The Way Forward", ed. A.J. Rugg-Gunn, Elsevier Applied Science, London, 1991, 163-168.

4. K. Dunnell and A. Cartwright, "Medicine Takers, Prescribers and Hoarders", Routledge and Kegan Paul, London, 1972.

5. J. Fry, D. Brooks and I. McColl, "NHS Data Book", MTP Press, Lancaster, 1984.

6. D. Brooks, Update, 1987, 35, 311.

7. G.W. Rylance, C.G. Woods, R.E. Cullen and M.E. Rylance, Br. Med. J., 1988, 297, 445.

8. M. Brandon and E.B. Sadler, Pharm. J., 1985, 234, 824.

9. E.B. Sadler, and M. Brandon Pharm. J., 1987, 238, 680.

10. E.B. Sadler and M. Brandon, Pharm. J., 1988, 241, 16.

11. I.C. Mackie, PhD Thesis, University of Manchester, 1991.

12. E.M. Bentley and I.C. Mackie, Pharm. J., 1993, 251, 606.

13. E.M. Bentley and I.C. Mackie, Health Ed. Research, 1993, 8, 519.

14. C. Appleby and I.C. Mackie, J. Inst. Health Educ., 1993, 31, 60.

15. I.C. Mackie and P. Hobson, Int. J. Paed. Dent., 1993, 3, 163.

Manufacturing Opportunities with Non-sugar Sweeteners

P. J. Sicard and Y. Le Bot

ROQUETTE FRERES, F-62136 LESTREM, FRANCE

1 INTRODUCTION

The origin of the consumption of non-carbohydrate sweeteners by man is linked to his habit of harvesting and eati fruits, mushrooms and algae. In fact, numerous fruits contain large amounts of polyols which contribute significantly to their sweetness (1, 5) (Table 1); Manna, the sweet exudate of Fraxinus ornus, mentioned in the bible as a providential food, has a high content of D-mannitol. Besides these bulk non-sugar sweeteners exists another group of natural intense sweeteners consisting of either proteins (miraculin, thaumatin, monellins) or complex glycosides (neohesperidin dihydrochalcone, stevioside).

TABLE 1.
Natural occurence of polyols

D-SORBITOL (3)		
Apples	2.6 - 9.2	g D-Sorbitol/l of juice
Pears	11.0 - 26.4	" " "
Cherries	14.7 - 21.3	" " "
Sour cherries	13.1 - 29.8	" " "
Plums	1.8 - 13.5	" " "
Rowan (S.aucuparia)	85	" " "
D-MANNITOL (5)		
Laminaria spp	10 %	of dry weight
Lactarius	15 - 20 %	" "
Agaricus	15 - 20 %	" "
XYLITOL (1)		
Yellow plums	0.93 %	of dry weight
Strawberries	0.36 %	" "
Cauliflower	0.30 %	" "

As is the case for a number of synthetic intense sweeteners, the discovery of the first member of this family of compounds was purely accidental. In 1878, REMSE and FAHLBERG (6) working on the oxidation of o-toluene-sulfonamide to synthesize o-sulfamoylbenzoic acid, unexpectedly obtained a condensed heterocyle characterized by an intense sweetness. The name "saccharin" was given to illustrate the exceptional properties of this new product.

For years, the production of saccharin remained limited (190 mt in 1900) and it was sold exclusively in pharmacies. However, due to sugar rationing during World War I, its consumption increased tremendously so that saccharin is still one of the leading intense sweeteners.

As concerns bulk non-sugar sweeteners, D-sorbitol started to be used as a sweetening agent for diabetics in the late 1920s (7). This application was mainly developed in Germany. In the USA, just after World War II, the first sugarless chewing-gums were produced, using a blend of crystalline sorbitol, liquid sorbitol and glycerin as a substitute for the traditional association between glucose syrup and crystalline sucrose.

In the early 1960s, a Swedish firm, LYCKEBY, developed a range of hydrogenated glucose syrups under the tradename "LYCASIN", with a view to producing non cariogenic hard-boiled candies (8). The tradename and patent rights were transferred to ROQUETTE FRERES in 1975. Then the technology for the production of LYCASIN® was modified so as to be in compliance with pH telemetry testing ; this resulted in a new product : LYCASIN® 80/55, presently the archetype of hydrogenated glucose syrups (9).

There exists now a wide range of intense and non intense sugar replacers which can be used alone or in admixture ; they can replace sucrose, glucose or fructose in most of their food applications.

All these products present specific properties that fit various types of utilization ; they are complementary to one another, no one being universal.

2 DIFFERENT TYPES OF NON-SUGAR SWEETENERS

Non-sugar sweeteners can be divided into two groups ; the first one encompasses the intense sweeteners, mainly synthetic molecules with the exception of the naturally occuring thaumatin ; the second one corresponds to bulk sweeteners, mainly polyols, either natural or synthetic.

Intense Sweeteners

As already mentioned, the discovery of some of the most important artificial sweeteners (saccharin, cyclamate, aspartame) was totally serendipitous.

Among the intense sweeteners presently authorized, most are obtained through chemical synthesis (10) (Figure 1). The main exception is thaumatin, a naturally occuring sweet protein extracted from the fruits of *Thaumatococcus daniellii* (11).

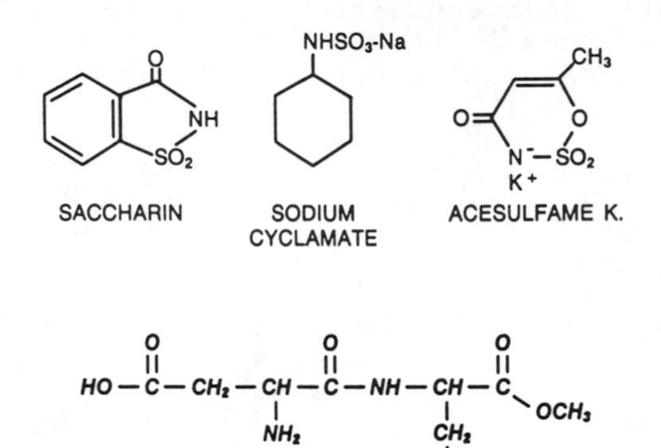

Figure 1 Main synthetic intense sweeteners

The sweetening power of these products varies widely (Table 2) as well as their physico-chemical characteristics.

Whereas acesulfame K, cyclamate and saccharin can be considered as very stable molecules, which is a definite advantage when they are involved in the preparation of food products undergoing thermal treatments, the stability of thaumatin and aspartame can be affected by temperature and pH.

Despite its very high sweetness thaumatin is considered more as a flavour enhancer or taste masker than as a sweetener (11).

TABLE 2.
Relative sweetness of the main intense sweeteners

SWEETENER	APPROXIMATE SWEETNESS
ACESULFAME K	200
ASPARTAME	180
SODIUM CYCLAMATE	30
SACCHARIN	300
THAUMATIN	2000

Besides these five well-known intense sweeteners, other products are presently under investigation by EC regulatory activities, in view of receiving limited or extended approval. Such is the case of neohesperidin, stevioside and sucralose.

Bulk Non-Sugar Sweeteners

As indicated, all the members of this group are poly-alcohols or polyols, either of natural origin or obtained through catalytic hydrogenation of simple or complex carbohydrates.

With the exception of some hydrogenated glucose syrups or maltitol syrups, all polyols utilized as bulk sweeteners can be cristallized as white odourless powders, more or less hygroscopic.

The sweetness of these products is generally below that of sucrose. Due to the reduction of the pseudo-aldehyde or ketone group of the corresponding carbohydrates they are much more stable and do not give rise to Maillard reactions.

From a biological point of view, all edible polyols can be considered as non-cariogenic; most of them are non insulinogenic.

D-Sorbitol. D-sorbitol or D-glucitol was first isolated in 1872 by J. BOUSSINGAULT, from the berries of the mountain ash tree, Sorbus aucuparia (12). It is naturally present in significant amounts in a variety of fruits (Table 1). However, its solubility in water makes a recovery from these natural sources impossible.

Historically, the first industrial production of D-sorbitol was obtained via the electrochemical reduction of D-glucose (13). Rapidly, however, it appeared that catalytic hydrogenation in the presence of Raney nickel (100-150 °C, 30-100 bars) was the technique of choice to obtain a product of quality.

Chemically, D-sorbitol is derived from D-glucose by the addition of two hydrogen atoms to the pseudo-aldehyde group at C_1. Thus, a raw material with a very high D-glucose content or better still pure D-glucose should be used for its manufacture.

Pratically, two possibilities may be considered :

a) The use of sucrose, which after inversion leads to an equimolar mixture of D-glucose and D-fructose :

$$1 \text{ sucrose} + 1\ H_2O \xrightarrow{\text{invertase}} 1 \text{ D-glucose} + 1 \text{ D-fructose}$$

The disadvantage of this raw material is that the hydrogenation of D-fructose leads to an equimolar mixture of D-sorbitol and D-mannitol (Figure 2), so that the hydrogenation of invert sugar yields 75 % D-sorbitol + 25 % D-mannitol, the latter being crystallizable in a pure state because it is much less water soluble than D-sorbitol. In this process, D-sorbitol is obtained as a mother liquor of D-mannitol crystallization and is contaminated by 9 % (W/W) of this polyol.

b) The use of pure D-glucose obtained by total hydrolysis of starch, followed by a purification by either crystallization or chromatography. In this case, pure D-sorbitol is obtained which can then be crystallized.

The physico-chemical properties of D-sorbitol (Table 3) reflect its strong affinity for water. Due to its ability to retain water or to release it into the environment according to the relative humidity, D-sorbitol is an excellent food stabiliser.

When crystallized, D-sorbitol shows polymorphism. Three distinct crystalline forms have been characterized (14). The alpha and beta forms are unstable and may be converted into the stable gamma form under certain temperature and humidity conditions ; this causes the sorbitol powder to cake when containing the three forms. Only the gamma form is stable.

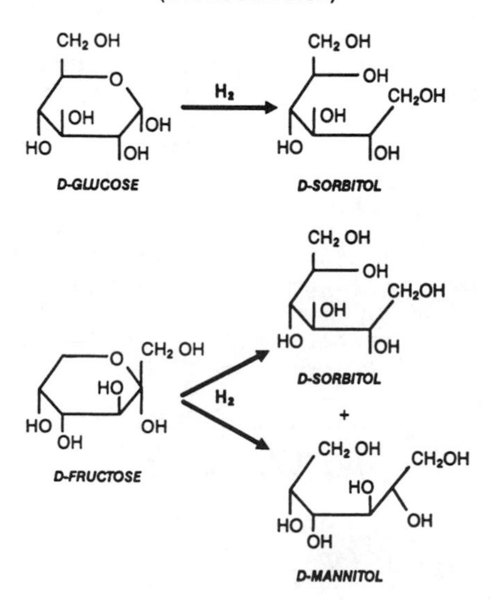

Figure 2 Sorbitol - Mannitol

TABLE 3.
Physico-chemical properties of D-sorbitol

CHEMICAL FORMULA	$C_6H_{14}O_6$
MOLECULAR WEIGHT	182.17
SOLUBILITY IN WATER AT 25 °C (g/100 g water)	235
HEAT OF SOLUTION AT 25 °C (cal/g)	-26.5
MELTING POINT (°C)	
α	88
β	94.5
γ	98.7
HEAT OF FUSION (cal/g)	43.45

D-Mannitol. D-mannitol or mannite is naturally present in seaweeds and mushrooms (Table 1). It was identified as the main constituent of manna in 1806 (15).

The extraction of *Laminaria digitata* has been used for a coupled production of alginate and D-mannitol. According to the season of harvesting, the upper part of this seaweed contains up to 10 % of polyol. This process is still currently used in China, but is of limited interest if we consider the amount of *Laminaria* to be processed to obtain significant quantities of D-mannitol (16).

Presently, more than 95 % of the D-mannitol produced in the world comes from the catalytic hydrogenation of D-fructose, either pure or in admixture with D-glucose (17).

The various sources of D-fructose commonly used are invert sugar, isomerized glucose syrup (HFCS), concentrated D-fructose solutions coming from the chromatography of invert sugar or HFCS and inulin hydrolyzate.

As indicated earlier (Figure 2), the catalytic reduction of D-fructose yields an equimolar mixture of D-mannitol and D-sorbitol from which it is easy to recover D-mannitol by crystallization, due to its much lower solubility in water.

The very low hygroscopicity and high chemical stability of D-mannitol make it an ideal product for the protection of labile molecules in dry formulations.

TABLE 4.
Physico-chemical properties of D-mannitol

CHEMICAL FORMULA	$C_6H_{14}O_6$
MOLECULAR WEIGHT	182.17
SOLUBILITY IN WATER AT 25 °C (g/100 g water)	22
HEAT OF SOLUTION AT 25 °C (cal/g)	-28.9
MELTING POINT (°C)	165-169

XYLITOL. Xylitol or xylite was first prepared from xylose obtained through an acid hydrolysis of wood chips by G. BERTRAND (18) and E. FISCHER (19).

On a commercial scale, it is manufactured by the chemical conversion of xylans, naturally occuring polymers of D-xylose (17, 20).

The most common sources of xylans presently used for the production of xylitol are listed in Table 5.

TABLE 5.
Natural sources of xylans of commercial interest

- Corn cobs (20-25 % D-xylose)
- Almond or hazel-nut shells (15-20 % D-xylose)
- Seed husks of Plantago sp. (46-72 % D-xylose)
- Pericarp of maize or rice kernels
- Cotton seed hulls
- Bagasse
- Birch wood chips
- Straw
- Waste liquors from paper and pulp industries

The hemicellulose fraction of these various raw materials which contains the xylans can be either separated by a preliminary treatment with alkali or directly hydrolyzed in the presence of cellulose. Hydrolysis is carried out under mild acid conditions, using H_2SO_4, H_3PO_4 or oxalic acid.

Following the process used, D-xylose can be purified by crystallization prior to its transformation into xylitol or hydrogenated without preliminary purification ; in the latter case, xylitol is purified by chromatography or crystallization.

As in the case of D-sorbitol and D-mannitol, hydrogenation is carried out under pressure, in the presence of Raney nickel (Figure 3).

XYLITOL

(HYDROGENATION)

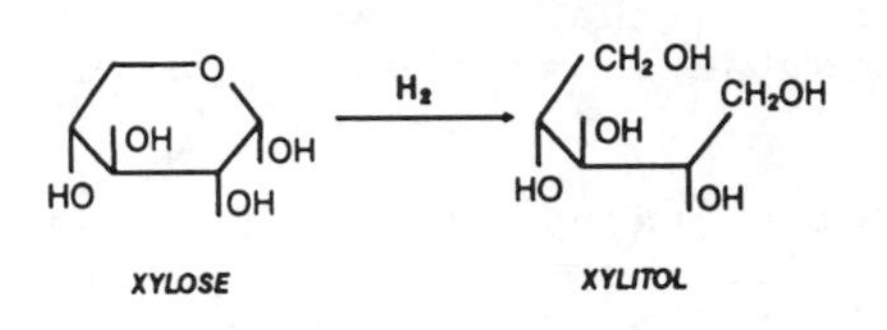

Figure 3 Catalytic hydrogenation of D-xylose to xylitol, under high pressure, in the presence of Raney nickel

TABLE 6.
Physico-chemical properties of xylitol

CHEMICAL FORMULA	$C_5H_{12}O_5$
MOLECULAR WEIGHT	152.15
SOLUBILITY IN WATER AT 25 °C (g/100 g water)	200
HEAT OF SOLUTION AT 25 °C (cal/g)	-36.61
MELTING POINT	
Metastable form	61-61.5
Stable form	93-95
BOILING POINT (760mm)	216

The main characteristic of xylitol is its sweetness, practically equivalent to that of sucrose, which enables its use without addition of intense sweeteners.

D-Maltitol. D-maltitol or (α (1 → 4)-glucosyl sorbitol), is a crystalline polyhydric alcohol which has been obtained rather recently (21) in a pure state, by hydrogenation of β-D-maltose which is the disaccharide produced by the action of β-amylase (EC 3.2.1.2) on liquefied starch.

Prior to this discovery, for a long time D-maltitol had been described as uncrystallizable. This situation was likely due to the presence of contaminants such as D-glucose, D-maltotriose and higher maltooligosaccharides besides D-maltose. Upon hydrogenation these carbohydrates yielded corresponding polyols, among which some were able to anticrystallize D-maltitol. The industrial production of D-maltose with a purity above 90 % being both difficult and costly, the problem of D-maltitol purity was solved when ROQUETTE FRERES (22) discovered that the chromatographic separation of reduced maltosaccharides was much more efficient than the separation of the maltosaccharides themselves.

Then, by chromatography of a syrup containing 50-70 % D-maltitol, it was possible to isolate a fraction containing 90-95 % of D-maltitol, from which this polyol crystallized very easily.

The transformation of β-D-maltose in D-maltitol is described in Figure 4.

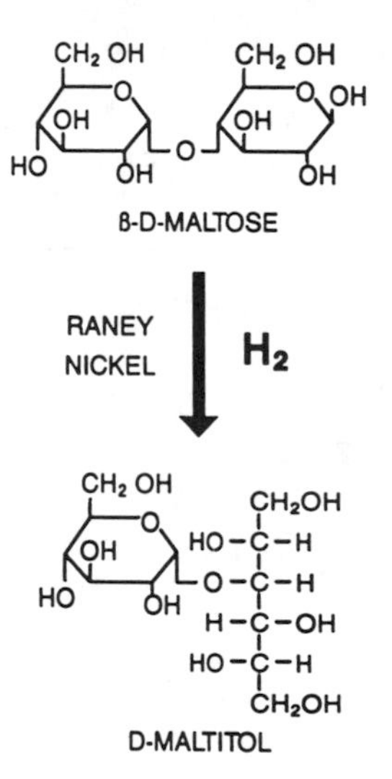

Figure 4 Hydrogenation of β-D-maltose to D-maltitol

High purity crystalline D-maltitol such as MALTISORB® (99 % pure) is probably the best substitute of sucrose, for technological and organoleptic reasons.

TABLE 7.
Physico-chemical properties of D-maltitol

CHEMICAL FORMULA	$C_{12}H_{24}O_{11}$
MOLECULAR WEIGHT	344.32
SOLUBILITY IN WATER AT 25 °C (g/100 g water)	167
HEAT OF SOLUTION AT 25 °C (cal/g)	-16.27
MELTING POINT (°C)	146.5-147

Hydrogenated Glucose Syrups - Maltitol Syrups. Since the original work of LYCKEBY (8) about the possibility of modifying advantageously glucose syrups by catalytic reduction, the family of hydrogenated glucose syrups (HGS) has undergone a very important development with as a main objective the production of non-cariogenic polyols utilizable in confectionery and gums.

The various types of products that can be obtained from starch by acid or enzymatic hydrolysis and their hydrogenated counter-parts are presented in Table 8.

TABLE 8.
Different types of complex polyols obtained from starch

Hydrolytic Agents[a] 1	2	3	4	5	Starch Derivatives	DE[b]	Corresponding Polyols[c]
	+				Maltodextrin	0-20	H L St
+					Acid glucose syrup	20-60	HGS
+		+			Acid-enzyme glucose syrup	20-60	HGS
	+	+			Maltose syrup	20-60	MS
	+	+		+	High maltose syrup	20-50	MS
+			+		Starch hydrolyzate	20-90	H St H
	+		+		Starch hydrolyzate	20-97	H St H

a) 1=Acid ; 2=α-amylase (EC 3.2.1.1.) ; 3=β-amylase (EC 3.2.1.2.) ; 4=Glucoamylase (EC 3.2.1.3.) ; 5=Debranching enzymes (EC 3.2.1.41 or 3.2.1.68).

b) DE=Dextrose equivalent

c) H L St=Hydrogenated Liquefied Starch ; HGS=Hydrogenated Glucose Syrup ; MS=Maltitol Syrup ; H St H=Hydrogenated Starch Hydrolyzate.

According to the type of catalyst used the hydrolysis of starch yields mixtures of reducing carbohydrates with various compositions. Acid degradation is random whereas enzymes act in a more specific way, especially glucoamylase that produces α-D-glucose, β-amylase that produces β-D-maltose and debranching enzymes who hydrolyze specifically the $\alpha(1 \rightarrow 6)$ bonds present in amylopectin, to give linear fragments.

As indicated in Figure 5, the hydrogenation of hydrolyzed starch, i.e. of a mixture of reducing saccharides (D-glucose, D-maltose, D-maltotriose and higher oligosaccharides) gives a mixture of the corresponding polyols (D-sorbitol, D-maltitol, D-maltotriitol and higher hydrogenated oligosaccharides).

The hydrogenation of saccharides resulting from starch hydrolysis leads to a significant modification of some of their physico-chemical and biological properties :

- Increased chemical stability,
- Increased affinity for water,
- Reduced tendency to crystallize,
- Increased chelating ability,
- Decreased susceptibility to enzymatic hydrolysis and fermentation,
- Change in the reactivity towards the body's hormonal system.

Except in the case of the reduction of D-maltose to D-maltitol, hydrogenation does not modify significantly sweetness.

The relative sweetness of D-maltitol is 95 % of that of sucrose; This along with the technological properties of D-maltitol has contributed to create a strong interest in maltitol and high-maltitol syrups family, for they can be used as such, without addition of intense sweeteners.

Historically, LYCASIN® 80/55 (9) was the first hydrogenated glucose syrup with a high maltitol content to appear on the market. Its composition (Table 9) had been devised so as to combine sweetness, technological properties and non-cariogenicity. More recently, products with higher maltitol concentration, such as MALTISORB® 75/75 (Table 9) have been created with a view to increasing sweetness of candies.

From a regulatory standpoint, initially all products resulting from the hydrogenation of hydrolyzed starch with a DE higher than 20 were identified as "Hydrogenated Glucose Syrups". The term "Maltitol Syrup" was defined by JECFA in 1988 for liquid products containing more than 50 % of D-maltitol on a dry weight basis (23). Besides their increasing use in food applications, maltitol syrups are now introduced in oral medicines due to their chemical stability and non-cariogenicity.

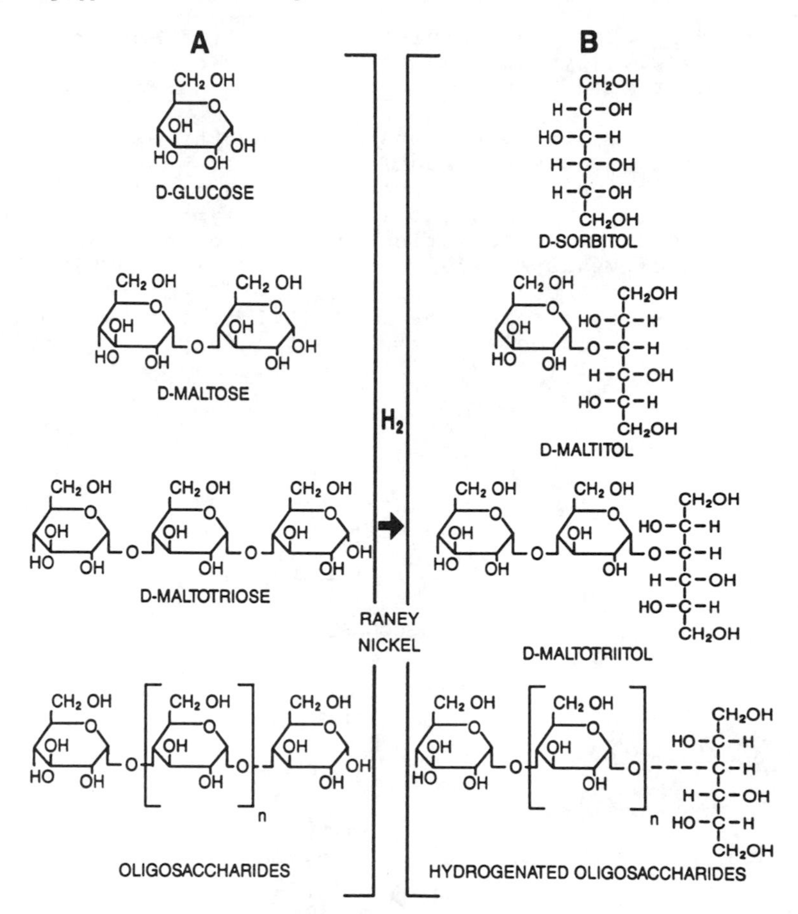

Figure 5 A) Chemical structure of mono-, di-, tri- and oligosaccharides resulting from starch hydrolysis. B) Chemical structure of corresponding polyols

TABLE 9.
Typical oligosaccharide composition (%) of Maltitol Syrups

	LYCASIN®80/55	MALTISORB® 75/75
DP1	3.0	1.5
DP2	54.4	75.5
DP3	19.3	14.0
DP4	0.8	1.5
DP5	2.8	1.5
DP6	3.5	0.5
DP7	2.9	1.0
DP8	2.3	2.0
DP9	1.0	2.0
DP10-20	8.9	2.0
DP sup. 20	1.1	3.0

DP=Degree of Polymerization, i.e. number of sorbitol and D-glucose sub-units entering into the composition of the oligosaccharide.

Isomalt. The discovery of isomaltulose, or palatinose, or 6-o-α-D-glucopyranosyl-D-fructose was achieved in 1952 in a fermentation broth of Leuconostoc mesenteroides (24).

In 1957, its presence was confirmed by WEIDENHAGEN and LORENZ in beet sugar manufacturing by-products (25). Then, the same authors discovered that a bacterium, Protaminobacter rubrum, contained an enzyme that was able to catalyze the transformation of sucrose into isomaltulose. Fifteen years later, the German firm SUDDEUTSCHE ZUCKER discovered and patented the transformation of isomaltulose into isomalt or PALATINIT®, a polyol which was in fact an equimolar mixture of glucosyl-sorbitol and glucosyl-mannitol (26, 27) (Figure 6).

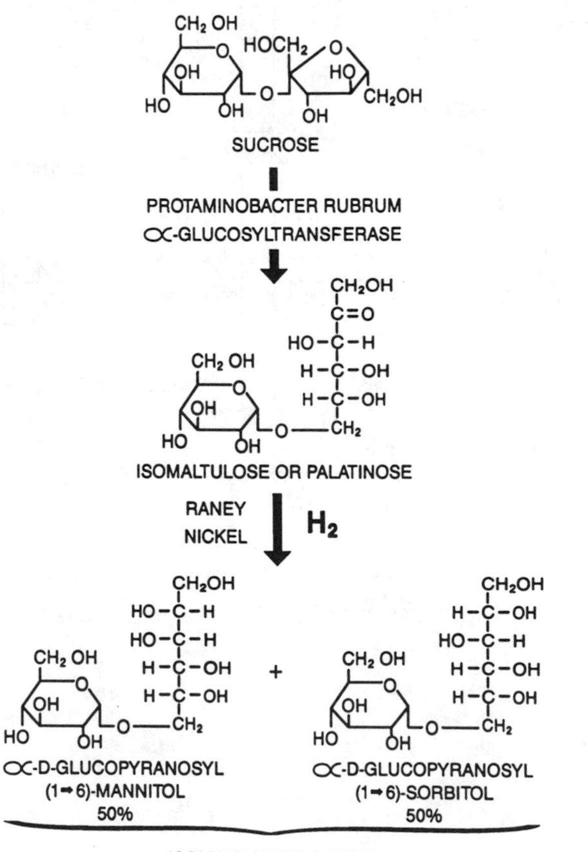

Figure 6 Enzymatic and chemical conversion of sucrose to isomalt

The properties of isomalt are listed in Table 10.

One of the main characteristics of isomalt is its low solubility in water, which can be attributed to glucosyl-mannitol (solubility in water at 25 °C expressed as g of product/100 g water is 83.5 for glucosyl-sorbitol and 25.2 for glucosyl-mannitol).

TABLE 10.

Physico-chemical properties of isomalt

CHEMICAL FORMULA	$C_{12}H_{24}O_{11}$
MOLECULAR WEIGHT	344.32
SOLUBILITY IN WATER AT 25 °C (g/100 g water)	39.6
HEAT OF SOLUTION IN WATER AT 25 °C (cal/g)	-9.4
MELTING POINT (°C)	145-150

Lactitol. Lactitol is a synthetic polyol which was obtained for the first time in 1920 by catalytic reduction of lactose (28).

It is presently produced by the Dutch firm CCA and by CULTOR by catalytic hydrogenation of high purity lactose (Figure 7).

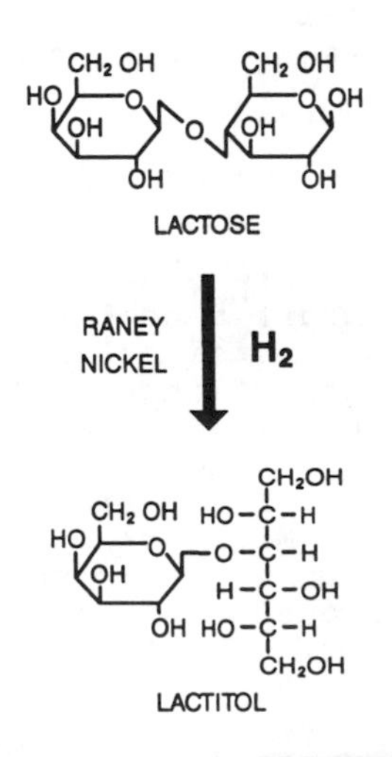

Figure 7 Hydrogenation of lactose to lactitol

Crystallization of lactitol, according to the conditions utilized, yields either a mono or a dihydrate.

The commercial product, LACTY®, is a monohydrate ; its properties are summarized in Table 11.

TABLE 11.
Physico-chemical properties of lactitol monohydrate

CHEMICAL FORMULA	$C_{12}H_{24}O_{11}H_2O$
MOLECULAR WEIGHT	362.33
SOLUBILITY IN WATER AT 25 °C (g/100 g water)	140
HEAT OF SOLUTION AT 25 °C (cal/g)	-13.9
MELTING POINT (°C)	70-80

Comparison of Polyols Physico-chemical and Technological Properties

Whereas intense sweeteners are only used to adjust the sweetness of finished products to the desired level, bulk sweeteners bring a number of valuable characteristics in terms of stability, texture, body, cooling effect, plasticity and, in the case of syrups, anticrystallization.

Solubility and Heat of Solution. The solubility of polyols in water varies strongly with the temperature, D-sorbitol being the most soluble. Whereas at 20 °C, the saturated solution contains 220 g of pure D-sorbitol/100 g water (69 % d.s.), at 40 °C the corresponding values will be 360 g or 78 % d.s.

Conversely, at 20 °C, mannitol and PALATINIT® have a low solubility in water.

Crystalline polyols are characterized by a negative heat of solution and impart a cooling effect in the mouth. The greatest effects are obtained with D-sorbitol and xylitol, whereas mannitol, although it has a higher heat of solution than D-sorbitol, is less cooling due to its low solubility.

Thermal Stability. Polyols are normally very thermostable. They do not give Maillard reactions. If a colouration appears during cooking, it is generally due to the presence of residual reducing sugars.

Viscosity of Syrups. Sorbitol and maltitol are available in the form of syrups. The viscosity of a sorbitol syrup at 70 % d.s. is lower than that of an equivalent sucrose solution. The viscosity of LYCASIN® 80/55 is higher than that of sucrose syrups and very close to that of a 60 DE glucose syrup.

Hygroscopicity. There is no relation between the hygroscopicity of a polyol in solution and that of the same in the crystalline state.

In solution, hygroscopicity depends on the molecular weight. Although sorbitol is an excellent humectant in solution, the crystalline form is not very hygroscopic particularly in the stable gamma form.

Mannitol has a very low hygroscopicity as it starts regaining moisture only at humidity higher than 95 %.

Physiological Properties of Polyols

Sweetness. Among all polyols, xylitol exhibits the highest sweetness, being practically as sweet as sucrose (Figure 8).

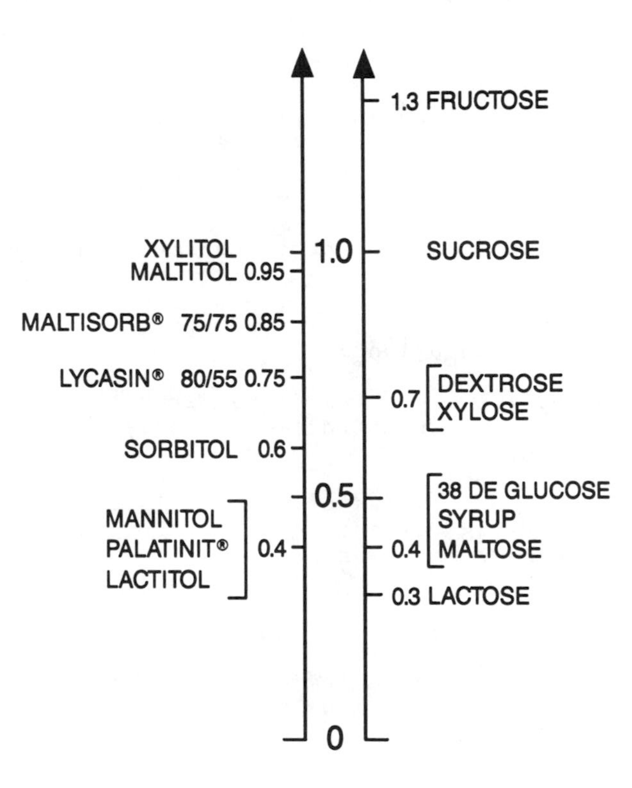

Figure 8 Relative sweetness of polyols and corresponding carbohydrates

The sweetness of maltitol is very close ; that of maltitol syrups depends on their richness in maltitol ; the sweetness of LYCASIN® 80/55 is 75 % of that of sucrose.

All other polyols are significantly less sweet, so that they generally have to be supplemented with intense sweeteners.

Metabolism. Polyols have to be divided into two families depending on whether they result from hydrogenation of monosaccharides or oligosaccharides.

Hydrogenated monosaccharides comprise xylitol, sorbitol and mannitol, whereas hydrogenated oligosaccharides consist of maltitol, lactitol, isomalt and maltitol syrups.

Two distinct biological mechanisms are involved in their assimilation.

a) The intestinal assimilation :

In contrast with D-glucose, which is absorbed by active transport, the hydrogenated monosaccharides are absorbed through the intestinal mucosa by an osmotic passive transport.

Except for D-mannitol which is incompletely metabolized in the organism and is partially excreted in urine, the hydrogenated monosaccharides which have passed through the intestinal barrier are wholly metabolized and provide 4 Cal/g.

The hydrogenated oligosaccharides must undergo enzymatic hydrolysis before passing through the intestinal mucosa.

b) The colonic fermentation :

In the colon, the microbial flora ferments polyols to volatile fatty acids. This fraction of polyols contributes to a caloric value of above 2 Cal/g.

Thus, the caloric value of polyols is comprised of between 2 and 4 Cal/g, depending on the balance between the two modes of assimilation.

In fact, the physiology of polyols can be influenced by complex parameters :

1. Absorption of polyols during or between meals,
2. Variable permeability of the intestinal mucosa,
3. Enzymatic activity of the mucosa towards the hydrogenated oligosaccharides,
4. Physiological status of the subject.

Considering the complexity of the phenomenon and the impossibility to allocate precise caloric values to individual polyols, the Scientific Experts of EC, in a wise decision, have agreed upon the fact that the mean energy value of those polyols would be 2.4 Cal/g (29).

Dental Properties. Dental caries is caused by the fermentation of sugars by the buccal flora. Polyols are not fermented by the buccal flora. This has been demonstrated by a variety of techniques, among which pH-telemetry occupies a place of choice (30, 31). A recent review of the dental properties of polyols by A.J. RUGG-GUNN has clearly demonstrated the beneficial aspects of their substitution of sucrose (32).

During the SAN ANTONIO Congress of the AMERICAN DENTAL ASSOCIATION in 1986, sorbitol was chosen as the official standard for non-cariogenicity testing.

3 INDUSTRIAL APPLICATIONS OF NON-SUGAR SWEETENERS

Food Industry

The main application of non sugar sweeteners in food is the production of sugarless confectionery.

A wide range of products can be produced in sugarfree form : hard-boiled sweets, chocolate, chewing-gum, directly compressed products, chewy sweets, jellies, hard gums, soft gums, wine gums, marshmallows, nougat, cream fillings.

Traditionally confectionery requires sugar and glucose syrup, e.g. a crystalline material and a syrup to control the crystallization (graining).

For sugarless confections, crystalline bulk sweeteners (sorbitol, mannitol, xylitol, maltitol, isomalt, lactitol) and LYCASIN®, which will replace glucose syrup and a part of sucrose, are available.

Directly compressed tablets. These are probably the easiest of sugarfree products to manufacture, although very specialized equipment is required for their production. In its simplest form flavour, colour, lubricant and sorbitol powder are mixed and then compressed into the desired shape. Lubricant is required to prevent the finished tablet from sticking in the press. Other crystalline polyols necessitate a granulation before being used for compression. Special directly compressible grades are available on the market.

Sugarless Chewing-gums. Standard chewing-gum may be conveniently divided into three phases. An insoluble gum base, a crystalline powder (sucrose) and a liquid phase (glucose syrup). Optimization of the phases gives a finished product which good appearance, texture, chewiness and shelflife. In sugarless chewing-gum, sorbitol, sorbitol/mannitol or sorbitol/xylitol can replace the sucrose and LYCASIN®, the glucose syrup.

Coatings. Sorbitol syrup and pure crystalline maltitol in solution can be used as direct replacements for sucrose in hard panning, although panning conditions have to be changed slightly to accomodate the products' unique characteristics.

LYCASIN® and sorbitol can be used in soft coating applications.

The thickness of the coating can be controlled in the same way as with sugared coatings and since the sorbitol crystallizes in the stable gamma form, the finished products are not very hygroscopic. The cooling effect of sorbitol is also obtained when the product is eaten.

Maltitol provides crunchiness and sweetness which are very close to those of sugar.

Hard-boiled sweets. Various polyols are useable with their pros and cons. The optimum choice depends on the process and type of packaging.

The use of hydrogenated monosaccharides, sorbitol, and xylitol, necessitate a depositing line whereas it is possible to use a forming "plastic line" with hydrogenated disaccharides or oligosaccharides.

The choice of the optimum polyol is then a function of the type of packaging.

In case of a moisture proof packaging, the simplest and cheapest solution consists in cooking maltitol syrup such as LYCASIN® to a temperature leaving a very low (1 %) final moisture content and to wrap candies just after forming or moulding.

When the packaging does not prevent totally any risk of water uptake, it is necessary to use together with maltitol syrup a crystallizable product which can crystallize on the surface of the candies thus preventing a viscous and tacky layer developing. Blends of LYCASIN®, with mannitol, isomalt or lactitol can be used.

Chocolate. Sucrose is the ideal sweetener for the production of traditional chocolate. The replacement of sucrose in the manufacture of sucrose free chocolate has proven difficult in the past because of the lack of a polyol that has the physical, chemical and organoleptic properties of sucrose.

Roquette's high purity crystalline maltitol provides the best properties for manufacturing of sugarless chocolate. Roquette's MALTISORB® high purity crystalline maltitol provides the properties of high sweetness, anhydrous crystalline form, low hygroscopicity, and high melting point.

These properties allow MALTISORB® crystalline maltitol to replace sucrose in a high quality chocolate coating using traditional manufacturing processes. For example, conventional refining and conching may be employed. High purity crystalline maltitol low hygroscopicity helps ensure remelted chocolate coating maintains its original physical properties.

Crystalline maltitol MALTISORB® provides the following advantages for chocolate manufacture :

- High thermal stability,
- Very low water content of the crystalline form,
- Low hygroscopicity,
- Small specific surface area.

During refining, the behavior of Roquette's crystalline maltitol is comparable to that of sugar, allowing the refiner to be set similar to standard chocolate making adjustments.

Crystalline MALTISORB® allows high-temperature conching which results in greater volatilization of undesirable flavor components (aldehyde and short chained fatty acids) and improved chocolate flavor development.

High temperature conching and fine grinding of the mass during the refining stage provides improved mass rheology (water evaporation, reduction and polishing of the maltitol crystals) and results in high quality chocolate.

The production of sugarless chocolate is possible with other polyols but necessitates low conching temperatures which is detrimental to the quality of finished products.

Jellies. To obtain clear jellies inhibition of crystallization is an absolute must. For this reason non crystallizable LYCASIN® is the optimum sweetener for the production of such confectionery items.

TABLE 12.
Polyols for sugarless confectionery

	TABLETS	CHEWING-GUMS	COATING	HARD CANDIES	CHOCOLATE	JELLIES
SORBITOL	+++	+++	++	+	+/-	+
MANNITOL	+	++	-	+	+/-	-
XYLITOL	+	++	++	+/-	+/-	+/-
MALTITOL	+	++	++	+	+++	+/-
LYCASIN®	-	++	+/-	++	-	++
ISOMALT	+/-	+	+/-	++	+/-	-
LACTITOL	+/-	+/-	+/-	+/-	+	-

The Table 12 summarizes the usage of various polyols in the above described applications.

Main Pharmaceutical Uses of Polyols

Traditionally, sucrose, lactose and starch have been used as excipients in pharmaceutical products.

The increased attention paid to oral hygiene has led to considering with much interest the possibility of replacing these cariogenic carbohydrates with bulk products that could bring the same technological advantages while being non-cariogenic.

The non cariogenicity of commercially available polyols being now a well established fact, they are increasingly used in various galenic forms, mainly for those drugs whose regular consumption could create a dental problem.

Presently, in the UK, products like sorbitol, mannitol, xylitol and LYCASIN® 80/55 are used in "sugarfree" products. These polyols are in compliance with USP XXII and BP 93.

Liquid non-crystallizable sorbitol (70 % d.s.) is used as a filler and sweetener in cough syrups, decongestant or antihistaminic syrups, multivitamins syrups and mouth washes.

Crystalline sorbitol is used as a filler, binder and sweetener in direct compression tablets, e.g. suckable tablets, where its hardness and cooling effect are appreciated. It can also be used as a filler in powder sachets, giving good flowability and rapid dissolution time.

Pyrogen-free sorbitol is used in large volume infusions.

Solid mannitol is available under two forms. As a powder, it is used in the manufacture of tablets or capsules (antacids, multivitamins complexes). In granular form, it can be used in direct compression.

Pyrogen-free mannitol has an application in large volume infusions.

Due to its dental properties, xylitol is appreciated as a filler and sweetener in gums, hard-boiled candies for coughs and mouth washes. LYCASIN® 80/55 is non crystallizable, non cariogenic and presents a good stability. Its main pharmaceutical applications are in cough syrups, paracetamol suspensions, and vitamin syrups. It can also be used for the preparation of hard-boiled candies.

4 EVOLUTION OF THE UTILIZATION OF POLYOLS IN GREAT BRITAIN

In 1983, MAFF authorized the utilization of a number of polyols in food applications : sorbitol, mannitol, xylitol, isomalt®, lactitol and HGS (containing from 50 % to 90 % maltitol). Surprisingly, pure crystalline maltitol was not included in this list.

Thus, we are presently awaiting the publication of the EC Directive on sweeteners, which could occur by the end of 1993, to put an end to this abnormal situation. This Directive will deal with the utilization of authorized sweeteners in food industry. Afterwards, we will have to wait for two more years for this Directive to be transposed to national regulations. Besides, a second EC Directive on specifications will also have to be published, but probably later on.

Up to now, the most successful areas for the utilization of polyols in Great Britain have been :

- Paediatric syrups, cough syrups, anti-histaminic syrups... for which more than 60 % of existing formulations are sugarless. This situation is a consequence of the recommendations made in the COMA Report n° 37 on "Dietary Sugars and Human Diseases (COMA : Committee on Medical Aspects of Food Policy). Now, practically all new liquid formulations are sugarless.

- Chewing-gums, 40 % of which are sugarless. In traditional confectionery, however, the use of sugar replacers is just starting.

What can reasonably be predicted for the next decade ?

In pharmaceutical industry, the situation is very clear, due to the positions taken by the health authorities.

In confectionery, we can anticipate an increased usage of non-carbohydrate sweeteners if the quality of commercial products justifies it and if the customers are well informed of the benefits resulting from a regular consumption of sugar-free products.

This could be extended to snacks, considering that the development of snacking habits is increasing consumers' dental risks.

Up to now, the countries where oral health is optimum (Swizerland, Finland) are also those where sugar-free products have been the most successful.

The diversity of available polyols obviously makes it possible to resolve all technological problems especially when crystalline maltitol is authorized, for this product is probably the best challenger of sucrose in very important applications such as chocolate.

5 CONCLUSION

The development of non-carbohydrate sweeteners, either intense or bulk has opened the way to new generations of food and pharmaceutical products, relying both on nutrition and health considerations. We can reasonably expect this trend to grow in importance during the coming years, consumers becoming increasingly health conscious.

Even if there seems to be a competition between these newcomers and traditional sugars, this consideration must be tempered by the fact that presently their market sizes differ by two orders in magnitude and also because most producers of bulk non carbohydrate sweeteners are equally producers of traditional carbohydrates.

REFERENCES

1. J. WASHUTTL, P. RIEDERER and E. BANCHER, J. Food Sci., 1973, 38, 1262.
2. British Food J., 1976, 78, 172.
3. K.C.F HEATON, F.D ROBINSON and M. LEWIN, Joint Meeting Midland Branch IFST and British Soc. Flavorist, Sutton Bonnington, 1980, 157.
4. R.E. WROLSTAD and R.E. SHALLENBERGER, J. Assoc. Off. Anal. Chem., 1981, 64, 91.
5. P.J. SICARD and P. LEROY, "Developments in Sweeteners", Vol. 2, Appl. Sci. Pub., London and N.Y., 1983, 1.
6. C. FAHLBERG and I. REMSEN, Berichte, 1879, 12, 469.
7. B.K. DWIVEDI, "Alternative Sweeteners", 2nd Ed., Marcel Dekker Inc., N.Y and Basel, 1991, 333.
8. LYCKEBY STARKELSE FORAOLING, AKTIE BOLAG, Fr Pat n° 1422060, 4 May 1964.
9. F. VERWAERDE, J.B LELEU and M. HUCHETTE, Fr Pat n° 2444839, to Roquette Freres, 11 Dec. 1978.
10. L. O'BRIEN NABORS and R.C GELARDI, "Handbook of Sweeteners", Blackie and Sons Ltd, Glasgow and London, 1991, 104.
11. J.D. HIGGINBOTHAM, "Alternative Sweeteners", Marcel Dekker Inc., N.Y., 1986, 103.
12. J. BOUSSINGAULT, Ann. Chim. Phys., 1872, 26, 376.
13. R.L. TAYLOR, Chem. Met. Eng., 1937, 44, 588.
14. S. QUINQUENET, "Contribution à l'étude du polymorphisme du sorbitol", Thèse UTC Compiègne, France, 1987.
15. L. PROUST, Ann. Chim. Phys., 1806, 57, 1.
16. N.A SOERENSEN and K. KRISTENSEN, US Pat n° 2516350, 1950.
17. P.J. SICARD, "Nutritive Sweeteners", Appl. Sci. Pub., London and N.Y., 1982, 145.

18. G. BERTRAND, Bull. Soc. Chim. Fr., 1891, 5, 555.
19. E. FISCHER and R. STAHEL, Ber., 1891, 24, 538.
20. C. AMINOFF, E. VANNINEN and T.E. DOTY, "Xylitol", Appl. Sci. Pub., London, 1978, 1.
21. H. MAMORU, H. HIROMI and M. TOSHIO, Jap. Pat n° 57134498 to Hayashibara, 19 Aug. 1982.
22. F. DEVOS and P.A GOUY, Fr Pat n° 2575180 to Roquette Freres, 20 Dec. 1984.
23. "Evaluation of certain Food Additives and Contaminants", 33rd Report of the Joint FAO/WHO Expert Committee on Food Additives, WHO Geneva, 1989, 19.
24. F.H. STODALA, H.J.H. DEPSELL and E.S. SHARPE, J. Am. Chem. Soc., 1952, 74, 3042.
25. R. WEIDENHAGEN and S. LORENZ, Z. Zuckerindust., 1957, 7, 533.
26. SUDDEUTSCHE ZUCKER AKTIENGESELLSCHAFT, Fr Pat n° 2179966, 11 April 1973.
27. SUDDEUTSCHE ZUCKER AKTIENGESELLSCHAFT, Fr Pat n° 2310354, 23 April 1976.
28. J.B. SENDERENS, Comptes-Rendus, 1920, 170, 47.
29. EC Council Directive 90/496/CEE on Nutritional Labelling, 23 Sept. 1990 (OJEC 6 Oct. 1990).
30. T. IMFELD, Heb. Odontol. Acta., 1977, 21, 1.
31. T. IMFELD and H.R. MUHLEMANN, J. Prevent. Dept., 1977, 4, 8.
32. A.J. RUGG-GUNN, "Nutrition and Dental Health", Oxford University Press, Oxford, 1993, 260.

Manufacture and Marketing of Sugar Free Confectionery

D. C. Pike

VIVIL UK LIMITED, UNIT C, BANDET WAY, THAME, OXFORDSHIRE OX9 3SJ, UK

My topic of lecture this morning is the manufacturing and marketing of Sugar Free Confectionery.

I was particularly pleased to receive the invitation from Professor Rugg-Gunn to address a distinguished audience; for me this is a unique occasion.

There is nothing fundamentally more important than acquiring a depth of knowledge within a specialised field, especially when each one of us has specific objectives within a common subject, yet viewed from completely different angles.

It is important therefore that we have an understanding of each others specialised fields, particularly if we have similar objectives. This is often a time-consuming process, demanding occasional looks over the horizon, ideally taken together, communicated to each other, and progressed by each other.

We are all familiar with precise research and development proposals and talk greatly about researching consumer habits and needs. However, changing these habits and needs that have been established over time, and doing so with a spirit of optimism can be most difficult and challenging. This is especially true of the Sugar Free consumer confectionery market development and Sugar Free Products in general.

Today, I would like to alert you or possibly remind you to some of the thinking behind the manufacture and marketing of Sugar Free sweets, as distinguished between Sugar Free Gum which will be covered by Adrian Piotrowski later today, with a major bias to the UK market when it comes to marketing. The views on marketing of Sugar Free confectionery are my own personal views. Each manufacturer will have its own marketing plan, even though our goals are similar.

Less than 10 years ago, the number and variety of non-sugar confectionery available to the consumer could be counted on one hand.

Today, as you can see, the consumer has a very wide and varied choice of products, from mints to jellies, soft-boiled and hard-boiled candies, chewing gum to chocolate.

In the last few years, here in the UK, Sugar Free sweets are just beginning to have impact on the consumer, yet remain a very small part of the total confectionery market.

These new product developments have not been brought about without considerable investment by large and small manufacturers.

Each one of these manufacturers has striven to produce a Sugar Free sweet that cannot be distinguished from a Sugar Based sweet.

In the late 60s, Sugar Free sweets were developed predominately based on Sorbitol. Most manufacturers were based in Switzerland and Germany, with the first sweets being Sorbitol Compressed Mints.

Sorbitol produced in fine powder form did not have to undergo cooking. By compressing this powder and adding peppermint flavouring, a consistent shape and taste for each sweet was produced.

Much experimenting with Sorbitol was conducted by manufacturers, and was consistently found to do little in enhancing the taste of the final product, especially when fruit juices were introduced.

In the early days of Sugar Free, restricted products and variety kept the market growth to a low level.

With further research, manufacturers were able to achieve better flavouring processes, and with the co-operation of raw material producers, Sorbitol Syrups were introduced and flavoured Sugar Free fruit sweets marketed.

Switzerland is considered by many the starting point for Sugar Free confectionery in Europe. Not only did manufacturers begin to market their products, but support from medical practitioners and dentists in the early days provided education for children and mothers. Today, Switzerland remains one of the highest consumers per capita of Sugar Free sweets.

The success in Switzerland led to a significant launch of Sugar Free confectionery in Germany, by the major sugar manufacturers. New varieties of products, strong packaging and a belief in the market opportunities for Sugar Free sweets by the leading manufacturers, puts Germany as the leading producer in total market volume for Sugar Free confectionery in Europe.

Prime manufacturers, along with prime producers of raw materials, worked very close together as the momentum within the consumer market was clearly growing. In May 1990, the German Ministry of Agriculture revised the law on Sugar Free bulk substances, making a much wider range of non sugar substances available to manufacturers. Sugar Free substances like MALTITOL, MANNITOL & ISOMALT provided even more products and even more choice for consumers.

From a manufacturer's view-point, even further advantages over Sorbitol were realised. In 1990, to enhance the sweetness of Sorbitol, high intensity sweeteners were used to bolster the sweetness level. Today these new substances almost achieve the same sweetness level of sugar, without additional sweeteners. By adding natural fruit, the final product cannot be distinguished between sugar alternatives.

All of this development in Europe has been fairly rapid. From the graph below, you can see that our European counterparts have significantly advanced Sugar Free confectionery.

SUGAR FREE MARKET IN EUROPE

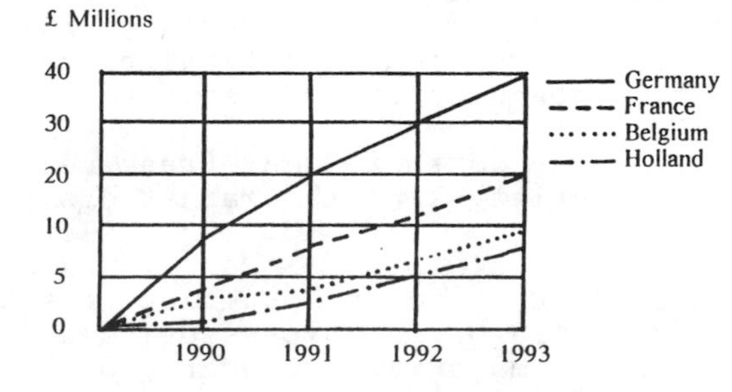

Population		UK estimated reduced calorie/ sugar free growth	
Germany	79.7 million		
France	55.9 million	1994 £11 million	)
Belgium	9.9 million	1995 £15 million	) at R.S.P.
Holland	14.8 million	1996 £20 million	)
UK	57.1 million		

Television advertising by all major brands, in Germany, continues to ensure that market growth will exceed the growth of sugar sweet sales into the year 2000.

Why is it then that with the progress in Europe the UK has failed to keep pace. My personal view is there are two different reasons:

Firstly: Unlike major German sugar sweet manufacturers, major UK sugar confectionery manufacturers have showed great reluctance to enter the market. The importance of the size of the UK sugar confectionery market, and the fact that the UK market is dominated by very few companies ensures protection of its sugar based market is paramount in their minds.

Secondly: Because of the size and domination of their brand names within the UK market, and in particular the eating habits of consumers, especially children, they consider that the laxative and flatulence effect of Sugar Free products will have a direct effect on their company and portfolio of products.

So before we can understand the opportunities for Sugar Free confectionery in the UK, it is important to understand the confectionery market and the consumer.

Lets take each point separately:

The current UK market for confectionery:

TOTAL CONFECTIONERY MARKET VOLUME AND VALUE

(percentage changes over 12 months)

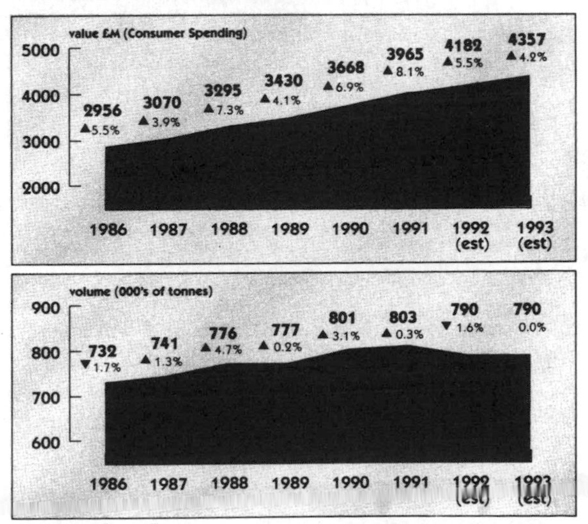

* For the seventh successvie year the amount spent by consumers in the UK on sweets has increased.

* The total confectionery market enjoyed an increase of over 5% in value of sales ahead of inflation.

* The average per capita expenditure per week on sweets rose by 5.1% to 41p.

* At the end of 1992 the UK confectionery market was worth in excess of £4.1 Billion.

<u>Sugar Sweets Market</u>

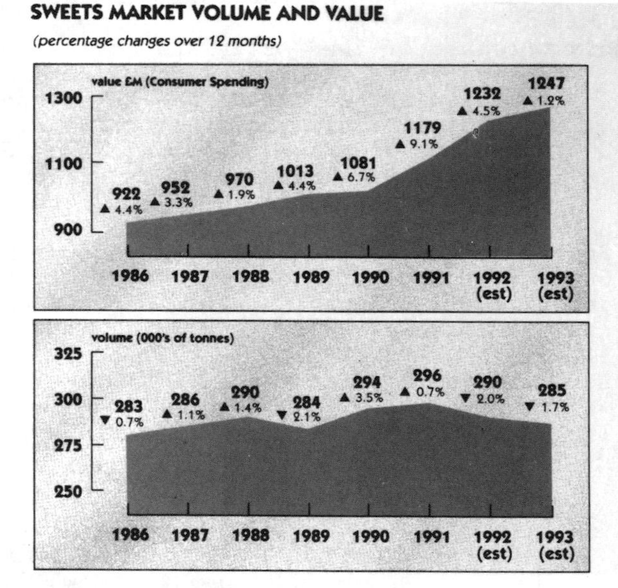

* The UK consumers annual consumption of sugar sweets per head is now 5.3 Kg.

* Worldwide that puts the UK in sixth place.

* Consumption of confectionery is bigger than the carbonated drink market.

* Also four times the size of the market for both ice cream and crisps.

Who buys what?

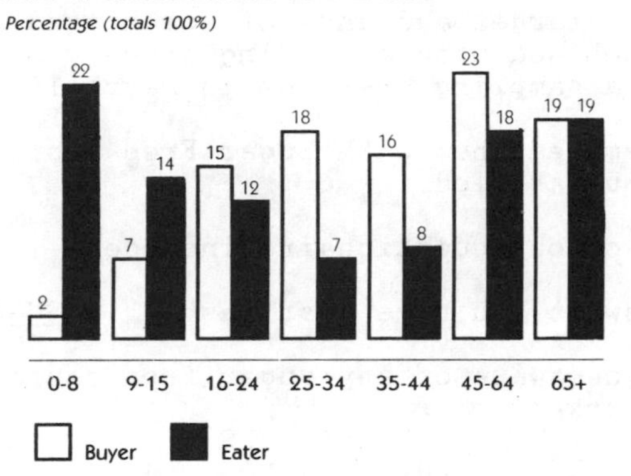

Here we can determine who is buying/eating.

Children buy 9% of all sweets but consume 36%.

Adults between the age of 25 and 44 buy 34% of all sweets but consume only 15%.

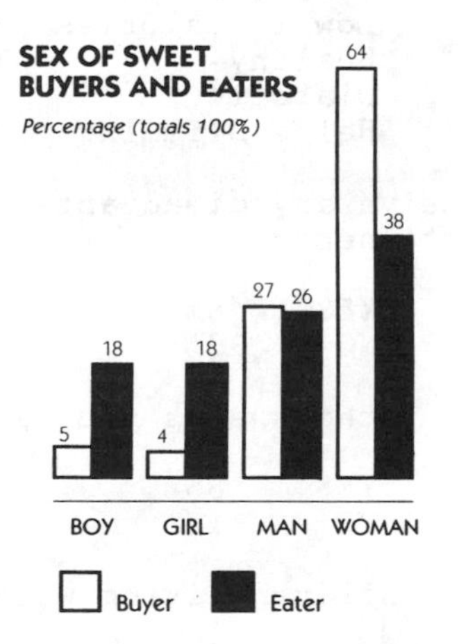

Twice as many women purchase sweets then men. 64% of women buy sweets yet consume only 38%.

These statistics are based on the year ending 1992 and show little change from the previous year.

As a company who's parent company are Germany's

leading manufacturers of Sugar Free confectionery, we further researched the UK confectionery market to identify the perception of consumers towards Sugar Free confectionery.

With a target audience of women aged between 18 and 54, combined with a leading grocery multiple we undertook a sampling exercise in March 1991.

We sampled them with Sugar Free confectionery which was unlabelled.

Part of our questionaire included:

Q3. The sweet you have just tasted is SUGAR FREE

Are you aware of any sugar free confectionery on the market?

60% said YES
40% said NO

Q4. Have you eaten Sugar Free confectionery before?

35% said YES
65% said NO

Q5. Please tick the box which best describes most accurately what "Sugar Free" means to you.

Safe for Teeth	21%
Low in calories	33%
Healthy	16%
Diabetic	6%
Helps Slimming	24%

Q6. Do you perceive any disadvantages of eating Sugar Free confectionery?

YES	8%
NO	92%

Q8. Would you buy the sweets you have just tasted?

YES	85%
NO	15%

Q9. Do you have children living with you?

YES	56%
NO	44%

Q10. Are you more likely to buy Sugar Free confectionery for:

Yourself	41%
Your children	39%

The whole family 10%
None of the above 10%

Q11. Do you worry about dental decay when you buy sweets or chocolate?

YES 87%
NO 13%

Q12. Are you aware of the molar "man" (Toothfriendly) symbol for confectionery?

YES 2%
NO 98%

Q13. Do you worry about the calorie content when you eat sweets or chocolate?

YES 89%
NO 11%

From these results and other consumer research that we entered into in 1991, and the growth that our European markets were making into the Sugar Free sweet market, we were convinced that the UK market would follow this trend.

It was obvious that availability of Sugar Free sweets to consumers was extremely limited. Although women were the main purchases of confectionery, suitable products to meet their lifestyle were unobtainable from their own source of purchase.

I reflect back to the symposium held here in 1990; it also happened to be the time when the EEC bodies were ratifying legislation on bulk sweeteners. I can well remember the excitement of Yves Le Bot and his colleagues from Roquette when news filtered through that all bulk Sugar Free substances would be reclassified in calorific terms; a 40% reduction compared to sugar.

I believe the statement "Healthier Eating Confectionery" was a by-product of the 1990 symposium.

Certainly following the launch of my own company's Sugar Free products, we tried to generate this new healthier eating image.

The Sugar Free market at the end of 1991 in the UK was worth only £6 million, 6 times smaller than the German market, and less than 1% of the total sugar market in the UK.

From our own confectionery survey we know the

majority of women are well aware of Sugar Free, and described what Sugar Free meant to them. From a manufacturers point of view, we know the key to success in meeting this need for healthier eating confectionery was to convince major supermarkets to re-think their position, give new exposure to Sugar Free products, and promote healthier eating confectionery generally.

We believed we had found a position for our brand. It had to be good, it had to have an identity. The consumer and retailer could see precisely what it was and where it belonged. We believed that if you have this all in place and make it crystal clear, then the market would come to you - not the other way round. I suppose it was a matter of know who you are and go for it.

By the end of 1992, distribution of Sugar Free confectionery showed tremendous improvement. (Figures do not include gum).

DISTRIBUTION OF REDUCED CALORIE AND SUGAR FREE IN UK AT DECEMBER 1992 (NIELSON)

32% Of all multiple grocers
40% Of all independent CTN's
9% Of all multiple CTN's
10% Of all independent grocers
8% Of all garages
1% Of all Co-ops

Our message to consumers was lifestyle.

Our promotional leaflets highlighted the key issues.

Sugar Free confectionery:

Helps prevent tooth decay
Helps prevent buildup of mouth bacteria
Has 40% fewer calories than normal sugar sweets
Helps prevent other sugar related diseases

It was important that we combined these statements with great tasting products. I was in a fortunate position to have my parent company's 15 years experience in the manufacturing of Sugar Free products, who provided me with great tasting products and informative packaging, developed from the German consumer knowledge gained by them.

Having now been involved in the marketing of Sugar

Free confectionery since 1990, I have been further encouraged with sales and distribution achieved.

Our UK market will have grown some 35% year on year. By the end of 1993 the Sugar Free sweet market will be worth over £11 million.

Consumers can now readily find a variety of Sugar Free sweets.

Effective TV promotion. I am sure you have seen two of Britain's most popular comedians promoting Sugar Free Clorets.

In February-March 93, my own company ran a TV campaign in Southern England, highlighting the reduced calorie benefits of Sugar Free sweets, and with further growth in our market, more TV spending will be made available.

We contrive to monitor consumer purchasing habits with women, who purchase Sugar Free sweets as much for themselves as well as their children.

New manufacturers have entered the market within the last few months, with more good tasting products enjoying increased distribution.

Major retailers as well as small confectionery shops prominently display Sugar Free products. 6 out of 10 small confectionery shops now stock at least one Sugar Free sweet. Some however still show reluctance: no doubt those next to schools that Dr. Adamson visited in her survey.

The consumer however remains the final arbiter. They have the decisive vote.

They exercise that vote with increasing perception. They are open to most opportunities in the market place and so we must inform them of the merits of a healthier eating confectionery option.

I mentioned two reasons for UK manufacturers' reluctance to enter the Sugar Free market.

The second point is the effects of over consumption of Sugar Free sweets and the laxative effect.

As distribution of Sugar Free sweets increases, so do the complaint letters of over consumption, especially in children.

It would appear that the statement carried on the majority of products, "Sorbitol if taken in excess can have a laxative effect, especially in young children", maybe not be enough for our Trading Standards and Environmental Health Department throughout the UK.

If sufficient data exists to provide a minimum/maximum dosage, especially for children, and as much as we have to promote the healthier aspects of Sugar Free we manufacturers remain susceptible to children over consuming product.

It is a fact that the majority of children handed a packet of sweets would normally consume these in less than 30 minutes. Small packs is one answer.

Major UK manufacturers for this reason have remained reluctant to market Sugar Free sweets, unlike German sugar producers who have enjoyed growth of Sugar Free products, yet retained their share of the sugar market.

From a European manufacturer's point of view, we can enjoy our position in the market place without the fierce competition of UK manufacturers. From a market development view, especially consumer awareness, growth of Sugar Free confectionery will continue to be inhibited whilst major UK manufacturers remain on the outside.

The problem of over consumption by children, and reactions of parents to the effects, should be considered fundamental to the education needed.

The most recent publication by the Toothfriendly Association is to be highly commended. Informative, educational and we must congratulate them. This is a major step forward.

Dr. Ian Mackie's regional attempt to have doctors prescribe and pharmacists sell SF OTC medicines has to be commended. This should become a national effort.

Doctors, Dentists, Dieticians & Manufacturers must all share a common interest in educating the consumer.

Our society in which we live expects an improving lifestyle, prevention rather than cure.

The benefits of our own particular field, in which we specialise, must justify the effort and produce the rewards expected by us.

Manufacture and Marketing of Non-sugar Chocolate

A. Zumbé, A. Lee, and D. M. Storey

DEPARTMENT OF BIOLOGICAL SCIENCES, UNIVERSITY OF SALFORD, SALFORD M5 4WT, UK

1 NUTRITIONAL GUIDELINES AND SUGARS

The most influential guideline in the last decade on nutrition and health was probably "A discussion paper on proposals for nutritional guidelines for health education in Britain"[1]. It defined nutritional targets and suggested ways of achieving them. Key recommendations for the population were specific reductions in intake of total fat, saturated fat, sucrose and salt and an increase in dietary fibre. Similar recommendations appeared in a later W.H.O. booklet[2] called "Healthy Nutrition".

The specific issue of sugar intake was tackled by the Committee on Medical Aspects of Food Policy in the COMA Report "Dietary Sugars and Human Disease"[3]. Sugars were categorised as "intrinsic" where they were part of the cells and "extrinsic" where they were free. For example, in an orange, sugars are intrinsic and in orange juice, extrinsic. Milk sugar, predominantly lactose, formed the third remaining category. The Committee concluded that tooth decay is positively related to the frequency and amount of non-milk extrinsic sugar consumption. Starchy foods, intrinsic and milk sugars have a negligible effect on tooth decay. It recommended that manufacturers produce low sugar or sugar-free alternatives to existing sugar-rich products, especially for children.

Standard sugar-containing milk chocolate will contain up to a maximum of 55% sugar (sucrose). Furthermore, consumers know that chocolate is rich in sugar and there is a common consumer perception that chocolate is bad for your teeth.

2 CHOCOLATE

Chocolate[4-6] can be defined as an almost anhydrous dispersion of fine, non-fat particles in a solidified fat phase.

Cocoa liquor, cocoa butter, sugar and milk are the four basic ingredients for making chocolate. By blending them in accordance with specific recipes, the three types of chocolate are obtained which form the basis of every product assortment, namely:

PLAIN CHOCOLATE:	cocoa liquor + cocoa butter + sugar
MILK CHOCOLATE:	cocoa liquor + cocoa butter + sugar + milk
WHITE CHOCOLATE:	cocoa butter + sugar + milk

Cocoa liquor: This is produced by roasting the raw cocoa beans, crushing them and grinding them into a fine paste. It has a dark brown colour and characteristic, strong odour which is indicative of the origin of the bean and the roasting conditions.

Cocoa butter: This is the pure oil which is obtained by pressing the liquor in hydraulic presses followed by filtration to remove the residual solid particles. The cocoa butter has important fuctions. It not only forms part of every recipe, but also gives the chocolate its fine structure, beautiful lustre and delicate, attractive glaze.

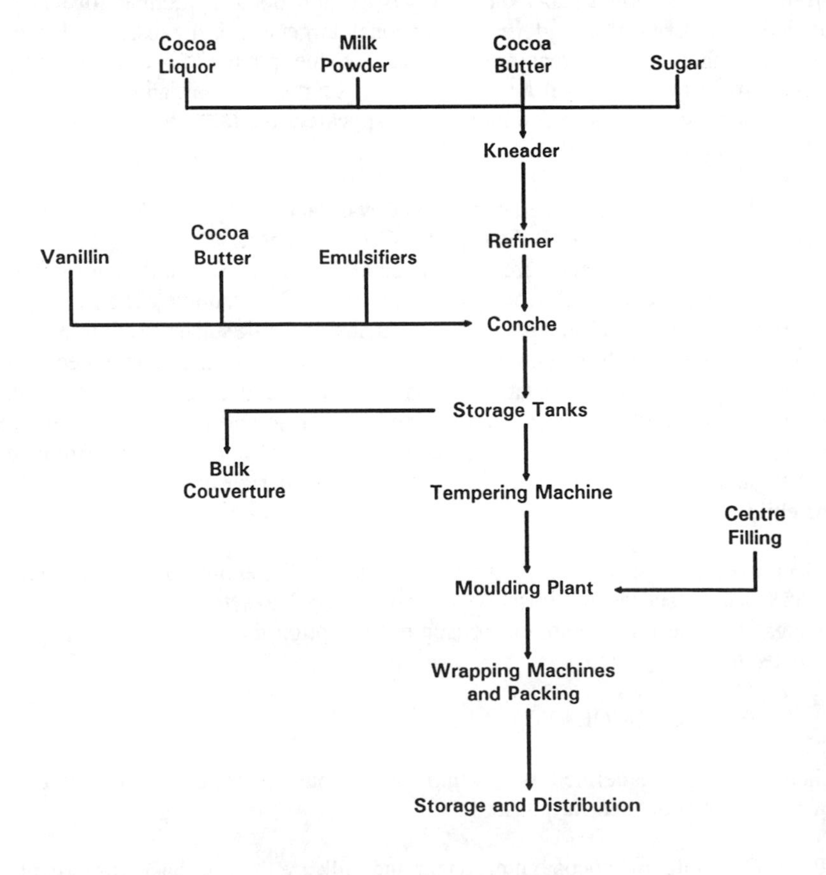

Figure 1 Production flow diagram; typical chocolate

A typical production flow diagram for the manufacture of chocolate is presented in Figure 1. The following steps are further explained:

Kneading

In the case of milk chocolate, for example, the cocoa paste, cocoa butter, milk powder, sugar and sometimes flavouring (e.g. vanilla) go into the mixer where they are pulverised and kneaded. The result is a homogeneous, paste-like mixture.

Refiner

Many designs of roller mills exist, although they are usually three or five vertically mounted steel rollers which rotate in opposite directions. Under heavy pressure they pulverise the tiny particles of cocoa and sugar down to a size of approximately 30 microns. The cloying paste is first forced between the two lowest rollers. Since the roller above is rotating at a faster speed it picks up the paste and feeds it upwards onto the next roller, until, with the pressure increasing all the time and the clearance between the rollers getting less and less, a very fine, flaky powder can be scraped off the top roller by a fixed blade.

Conching

The chocolate paste is still not smooth enough to satisfy our palates. The separate flavours of the individual ingredients have not yet combined; the pure, rounded flavour is still missing.

Conching is a process by which the chocolate paste is heated, usually to about 80°C, and constantly stirred for hours to give a velvet smoothness. Additional cocoa butter, vanilla and emulsifier (for example lecithin) are introduced into the mix. A kind of aeration of the liquid chocolate paste takes place in the conche, the bitter taste gradually disappears and the flavour is fully developed. At the same time, the ultimate homogeneity of the chocolate is produced and a soft film of cocoa butter begins to form around each of the extremely small particles. The chocolate is no longer sandy but dissolves meltingly on the tongue.

Storage

Chocolate can be stored in its liquid state or for prolonged storage is solidified in large blocks. The blocks can be reheated and liquified again.

Tempering

Before the forming process, the chocolate paste must be heated to 50°C and then cooled to a specific temperature a little over 30°C depending on the product. This thickens the chocolate and imparts the right flow properties for filling the moulds. This complex operation is performed in the tempering plant and is necessary to give the final chocolate product a delicate composition and a uniform structure. Essentially, tempering is a method of inducing cocoa butter to crystallise in a stable form in the fluid chocolate mass. This process is necessary to ensure a long shelf life; incorrectly tempered chocolate has a short shelf life, poor gloss and inadequate stability. A fault named 'bloom' can develop. Fat bloom is the description given to chocolate with a white sheen or, sometimes, individual blobs on the surface of the chocolate.

The purpose of each processing step is summarised in Table 1. This is of course an over-simplified description of the process and reference to specialised literature is recommended.

Table 1 The manufacture of chocolate and the purpose for each manufacturing step

Standard Procedure	Goals
Kneading	Mixing the ingredients
Refining	Grinding. Reducing the size of the sweetening crystals
Conching (wet/dry)	Changing the flavour Improving the rheology Eliminating the water
Tempering	Seeding the cocoa butter to obtain the stable crystalline form
Moulding	Solidification of the mass, final cocoa-butter crystallisation
Packaging	Presentation, preservation, transport

3 STANDARD OF IDENTITY OF CHOCOLATE

The Cocoa and Chocolate Product Regulations both at the U.K.[7] and E.E.C.[8] level precisely define the definitions and compositional requirements for a whole range of chocolate products as shown in Table 2. The regulations also impose limits on additives and other ingredients that may be used in chocolate products. Labelling requirements and restrictions for chocolate products and sizes in which certain chocolate products, in bar or tablet form, must be sold are also prescribed. The reserved description must be applied in trade and is, therefore, the name of the food for these products.

In regard to chocolate, "chocolate" is any product obtained from cocoa nib, cocoa mass, cocoa, fat-reduced cocoa, or any combination of these ingredients and sucrose, with or without added extracted cocoa butter and containing a minimum of 35% total cocoa solids, including a minimum of 14% dry non-fat cocoa solids and a minimum of 18% cocoa butters.

Alternative ingredients to replace the sugar, for example sugar alcohols and intense sweetening agents, are not permitted under the chocolate standard of identity. Such products are usually described as "chocolate flavour" bars. However, the food legislation is in a state of flux and a modification in the regulation to include this category of products as chocolate is probable.

4 U.K./E.E.C. NUTRITIONAL LABELLING GUIDELINES FOR SUGAR-CONTAINING FOODS

Current U.K. requirements relating to "sugar-free" require that 100g/100ml of the foodstuff contains no more than 0.2g sugars, although it is to be expected that this will shortly be brought into line with the E.E.C. claims directive of 0.5%. There is discussion to move it to 1.0%[9,10].

The U.K. requirement relating to "reduced" in sugar requires that the sugar component is not more than three-quarters of that in a similar food for which the claim is not made. In the E.E.C. claims directive, similar conditions are specified.

The U.K. labelling regulations, "low" is defined as not more than 5g of sugar per normal serving or per 100g/100ml, whichever is the higher.

The E.E.C. Council directive on sweeteners for use in food-stuffs defines "no added sugar" as without added mono- or disaccharides or any other food- stuff used for its sweetening properties.

Table 2 Standard of identity of milk chocolate

	Min % dry cocoa solids	Min % non-fat cocoa solids	Min % dry milk solids	Min % milk fat	Min % total fat	Max % sugar
Milk chocolate	25	2.5	14	3.5	25	55
Quality milk chocolate	30	2.5	18	4.5	25	50
Cooking milk chocolate	20	2.5	20	5	25	55
Milk vermicelli/flake	20	2.5	12	12	25	66
Gianduja nut milk*	25	2.5	10	3.5	25	55
Couverture milk	25	2.5	14	3.5	31	55
White chocolate**			14	3.5		55

* + min 15% - max 40% hazelnut paste
** + min 20% cocoa butter

Aromas that have a chocolate or milk fat taste are not allowed

5 NUTRITIONAL COMPOSITION OF CHOCOLATE AND CARIOGENICITY

The nutritional value of standard chocolate and the carbohydrate composition of typical chocolate are shown in Tables 3 and 4.

Table 3 Nutritional composition of chocolate[4]

Component	Plain	Per 100g Milk	White
Protein	3.2 g	7.6 g	7.5 g
Fat	33.5g	32.3g	37.0g
Carbohydrate	60.3g	57.0g	52.0g
Pure lecithin	0.3 g	0.3 g	0.3 g
Theobromine	0.6 g	0.2 g	--
Mineral substances			
Ca	20 mg	220mg	250mg
Mg	80 mg	50 mg	30 mg
P	130mg	210mg	200mg
Trace Elements			
Fe	2 mg	0.8mg	traces
Cu	0.7 mg	0.4mg	traces
Vitamins			
A	40 IU	300 IU	220 IU
B1	0.06mg	0.1mg	0.1mg
B2	0.06mg	0.3mg	0.4mg
C	1.14mg	3 mg	3 mg
D	50 IU	70 IU	15 IU
E	2.4 mg	1.2 mg	traces
Energy			
kJ	2080	2160	2260
kcal	495	515	540

Standard nutritional data used for the carbohydrate containing ingredients:

Ingredient	Moisture	Protein	Fat	Carbohydrate %
Wholemilk powder	3.0	26.4	26	39.5 (lactose)
Skimmilk powder	3.0	1.3	1.3	53.4 (lactose)
Cocoa liquor	2.0	12.0	55.0	10.0
Sugar				100 (sucrose)

Chocolate is rich in fermentable carbohydrate, notably added sugar (sucrose) and milk sugar (lactose) which can be fermented by the oral bacteria to form acid. It is no surprise therefore that standard sugar-containing chocolates will fail the plaque pH telemetric test for the determination of "tooth-friendly" chocolate as defined by the Swiss Tooth-Friendly Association[11-14].

However, unlike many manufactured foods (for example sugar confectionery) and beverages (for example soft drinks) which are rich in added sugar, it should be stated that chocolate has a number of properties which may reduce the cariogenic potential of the sugar in the chocolate.

Chocolate is rich in minerals, namely calcium, phosphorus and magnesium which are important for teeth. The cocoa liquor is rich in tannins and polyphenols which are known to have bacteriostatic properties and this point in particular merits further research. The fat content of chocolate is also thought to create a thin film over the tooth surface thus contributing towards a protective layer. The milk proteins in chocolate also have a certain buffering potential and this will reduce some of the acidity which is created when the sugar component is fermented. A recent patent even describes the presence of free fluoride in cane sugar that could be used for the manufacture of chocolate[15].

Table 4 Typical carbohydrate composition in chocolate

	Starch	Milk sugar	Non-milk sugar	
			Intrinsic	Extrinsic
Sugar Chocolate				
White chocolate		14.3		43.9
Milk chocolate	1.5	9.2		48.9
Non-Sugar Chocolate*				
White chocolate		14.3		
Milk chocolate	1.5	9.2		

* sugar ingredient replaced by bulking agents

Chocolate could be made much less cariogenic by introducing various factors into the ingredient mix, for example by increasing the buffering potential (by the use of urea or other salts) or by the controlled fortification with a natural ingredient which is a source of free fluoride. Unfortunately, this field of applied research has not been pursued to its full potential, although it is to be expected that such products will emerge in due course. Incidentally, the manner in which confectionery is eaten can also influence the cariogenic potential. Chewing, for example, can actually promote saliva secretion which will in turn help neutralise the acidity.

6 THE MANUFACTURE OF NON-SUGAR CHOCOLATE

Essentially, the sugar ingredient is replaced by a combination of a sugar alcohol and an intense sweetener. The sugar alcohol provides the bulk and some of the sweetness whilst an intense sweetener is used to top up the sweetness level. The level of sweetness of these individual ingredients as compared to sugar (sucrose) is shown in Table 5[16-18].

Table 5 Sweetness level of different alternative sweeteners compared to sucrose (sucrose = 1.0)

Sweetener	Sweetness
Bulk Sweeteners	
Sorbitol	0.6
Mannitol	0.5
Xylitol	1.0
Lactitol	0.4
Maltitol	0.9
Isomalt	0.5
Maltitol syrup	0.6 - 0.8
Polydextrose	0
Intense Sweeteners	
Saccharin	400
Cyclamates	40
Aspartame	200
Acesulfame-K	200
Sucralose	600

The main problem is that many of the sugar alcohols used in chocolate making are crystalline in the hydrated form. If the temperature is too high during refining and conching, this may result in their water of crystallisation being released and the chocolate's mass solidifying. The temperature on the rollers of the refiner and in the conche is critical and must be carefully monitored. An exception is for maltitol which is available in an anhydrous form and there is thus no risk of water of crystallisation being released. In reality, an experienced food technologist is aware of this issue and his choice of sugar alcohol will not be limited to this particular phenomenon.

Sorbitol belongs to the first generation of non-sugar chocolates and has largely been abandoned due to the poor taste characteristics in chocolate and to the relatively high laxative potential.

In general, sugar alcohols create a cooling effect on the palate and this is a function of the heat of solution and the rate of dissolution. It is not generally desired in chocolate. Xylitol and sorbitol have the highest cooling effect at 37°C.

The choice of sugar alcohol for use in chocolate will generally fall on either isomalt, lactitol or maltitol, with the deciding criteria usually depending on a combination of the following factors:

- price.
- hardness of the crystal, as this will influence the working life of the rollers in the refiners.
- the achievable size profile and shape characteristics of the refined crystals in the chocolate matrix. This is a crucial parameter for limiting the total fat which is needed to obtain the desired viscosity.
- hygroscopicity of the sugar alcohol. Moisture absorption by the bulk replacement ingredient during storage or manufacture must be avoided because free water will greatly increase the chocolate viscosity. High viscous chocolates are difficult to mould and will detrimentally influence the melt properties of the chocolate. In addition it is thought that excess free water will influence the taste characteristics of the chocolate in storage. Sorbitol is especially unsuitable because of its relatively high hygroscopicity (Figure 2).
- water content of the sugar alcohol crystal.
- thermal stability of the sugar alcohol. It is indispensable to avoid the crystals melting during the chocolate manufacture. Should the crystals melt they will agglomerate and this will increase the chocolate mass viscosity. The most sensitive points to regulate and monitor are the refining and conching steps.
- organoleptic quality of the finished chocolate. The distinguishing characteristics will be low cooling effect, lack of astringency and texture. Of course some sugar alcohols are sweeter than others, but this is easily compensated for by the addition of an intense sweetener.
- required fat content to achieve a similar viscosity to a sugar-containing chocolate.
- the appropriate conching temperature for a given sugar alcohol, as the conching temperature is an important factor for achieving the final desired taste.

A number of intense sweeteners are available. The choice will largely depend on the sweetener or combination of sweeteners that give the chocolate an overall taste profile similar to standard sugar-containing chocolate. Of the sweeteners currently permitted, the choice usually falls upon Aspartame or Acesulfame K.

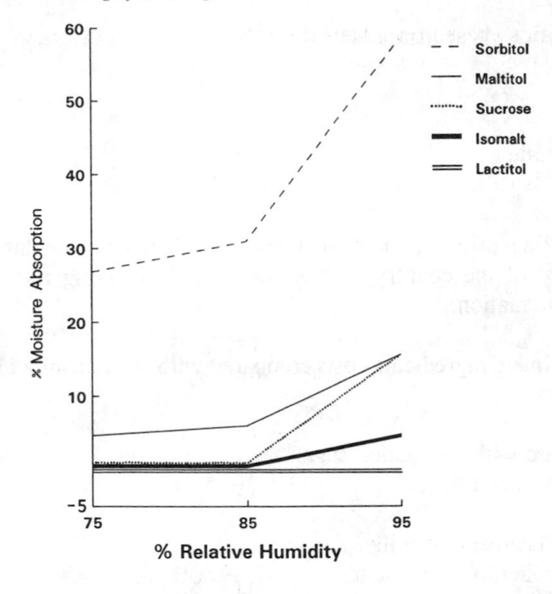

Figure 2 Hygroscopicity. Moisture absorption, 7 days at 25°C

Acesulfame K is extremely stable in chocolate whilst aspartame is especially sensitive to high temperature. This issue can be overcome by the addition of the aspartame to the chocolate at the end of conching, when the chocolate liquid mass is at a lower temperature.

It is advisable to manufacture non-sugar chocolate on dedicated production lines and not on equipment usually used for standard chocolate, otherwise the non-sugar chocolate will contain residual sugar. Experience has shown that it is extremely difficult and time-consuming, not to mention the waste chocolate that accumulates, to entirely eliminate the standard chocolate. The tempering equipment and connecting pipe work along the line are especially difficult to clean out.

7 PRICING

The current prices of bulk sweeteners and intense sweeteners are listed as follows:

Approximate prices based on £ sterling/ton*.

Sugar sucrose	620
Fructose	1100
Sorbitol	1000-1300
Xylitol	3900
Mannitol	2300
Isomalt	2100
Lactitol	2040
Maltitol (crystalline)	2300

Approximate prices based on £ sterling/kg*.

Aspartame	50-70
Acesulfame	55-65
Sodium Cyclamate	3-4
Saccharin	2-3

* the sterling price of many of these ingredients is calculated from the currency of the country of manufacture, thus giving rise to a certain price fluctuation.

Approximate ingredients costs compared with regular milk chocolate are listed as follows:

A milk chocolate with "no added sugar" (with lactitol or isomalt)	+ 55%
A sucrose-free/lactose-free milk chocolate (with lactitol or isomalt)	+ 60%

The final retail price will of course be influenced by a number of factors, but is clear from the ingredient costs that price parity to regular chocolate is simply not possible. In general, reduced sugar milk chocolate current products positioned for the mass market tend to carry a retail premium value (on an equivalent weight basis) of 50-250%. Non-sugar chocolates positioned for diabetics usually tend to have the highest retail prices, presumably due to the limited market volume, the distribution channel and the point of sale in pharmacies and dietetic stores where the trade is accustomed to higher margins.

Price, therefore, is a major factor which will inhibit sales in main stream outlets. Consumers are undoubtedly interested by sugar-reduced and sugar-free claims, but there is a limit to how much extra they are prepared to pay. A case in point is sugar-free chewing gum where the premium price is much less and the sugar-free market share in this segment is substantial.

The authors predict that price margin will continue to fall over the next few years and a retail price differential of approximately 25% will be achievable. This will be due to the fierce competition between the sugar alcohol suppliers with a downward trend for the ingredient price and also due to competition between manufacturers who will josle for market dominance in this growing segment.

8 NON-SUGAR CHOCOLATE AND SATIETY

The chocolate recipes discussed in this paper are rich in fat (from ca. 30-40% by weight) and contain intense sweeteners to make up for the sweet intensity lost by the omission of added sugar (sucrose). For comparison, a standard sugar-containing milk chocolate will contain ca. 45-55% of sugar by weight; a "no added sugar" milk chocolate recipe will contain about 9% and a truly sugar-free recipe will contain less than 0.5%. The calorific content of the reduced sugar and sugar-free recipes, with polyol as the sole replacement bulking agent will contain about 10% less calories than the standard chocolates with an equivalent fat content. Recipes which contain a combination of sugar alcohol, polydextrose or oligofructose and a similar fat content contain up to about 20% less calories.

In non-sugar chocolates, the reduction of total energy content is not so great, although the glycaemic response will differ significantly[19]. However, the issue of whether hedonic responsiveness is directly tied to some index of glucose or lipid metabolism remains unresolved.

In the few studies in which foods have been used to study sweetness, hedonism and satiety, it is clear that, during actual consumption, sweet sugar-reduced and sugar-free foods with non-glucose dependent sweeteners can be as satisfying as sugar-containing foods. Whether there is subsequently some craving and compensation for the missing carbohydrate is an open question[20,21].

As chocolate tablets and count-lines are nutritious snack foods, regularly eaten between meals to provide some relief from hunger and often as an integral part of a meal taken out of the home, experimentation is required to throw light upon the relative importance of the pleasantness of energy-rich non-sugar sweet foods and perceived hunger in the short- and long-term control of food intake. Such data are virtually non-existent.

9 GASTROINTESTINAL TOLERANCE OF SUGAR ALCOHOLS

Sugar alcohols (polyols) are safe for use in foodstuffs and are permitted food ingredients. However, it is well known that they may induce mild side effects when consumed in excess. Typical intolerance symptoms are described in Table 6.

Such symptoms are equally common in the consumption of a range of different foods, notably for example cabbage, beans and pulses, onions, pears, cherries, plums and peaches. These gastrointestinal symptoms are therefore not a phenomenon specific to polyols. Whenever there is a relatively high proportion of low digestible carbohydrates in the diet, the corresponding reaction of the gastrointestinal tract must be reckoned with. A large individual variance is known regarding gastrointestinal tolerance. Some people can tolerate larger amounts of "critial food" without problems than others.

Table 6 Intolerance Symptoms

Symptoms	Description	Cause
Laxation (least severe)	Increased stool frequency	Osmotic effect of intact polyol
Diarrhoea	Loose or watery stools	Osmotic effect of intact polyol
Colic	Lower abdominal pain or discomfort	Excess intestinal gas due to fermentation of polyol
Stomach ache	Upper abdominal pain or discomfort	Excess intestinal gas due to fermentation of polyol
Wind	Flatus	Excess intestinal gas due to fermentation of polyol
Bloating	Abdominal distention	Excess intestinal gas due to fermention of polyol

The dominant factors influencing tolerance are:

- The usual diet (the level of consumption of low digestible carbohydrates).
- The form of intake (in solid form or in solution, either in isolation or as an ingredient in a food product).
- The rate and total quantity consumed per day by an individual.
- Whether consumption is on an empty stomach or after a meal.
- Adaptation of the individual flora.

Emptying of the stomach and the quantity of poorly digested carbohydrates in the colon and the small intestine are influenced by these factors.

The Anatomical and Physiological Background

Stomach. When food is eaten fluid is either added or removed in the stomach depending upon the consistency of the material consumed. Many drinks virtually pass right through the stomach and very quickly reach the small intestine. Fat is known to retard gastric emptying and thus chocolate (which contains about 30% fat) is an ideal food for the consumption of polyols.

Small intestine. Secretion of enzymes would normally break down the carbohydrate including disaccharides (e.g. sugar sucrose and milk sugar lactose) into an absorbable form. However, polyols are not or are only minimally hydrolysed by enzymes in the small intestine whereas sucrose and lactose are easily hydrolysed into sugar monosaccharides and absorbed by the time they reach the upper section of the small intestine.

Every molecule in the small intestine binds a certain quantity of water (osmotic effect). The osmotic pressure of the water-soluble compounds depends on the number of molecules in the solution. Based on an equivalent gram per litre solution, the osmotic pressure of water-soluble components is higher the lower the molecular weight. Monosaccharides, therefore, have a higher osmotic pressure in solution than disaccharides.

Large intestine. In the large intestine, there is water absorption and the food residues become more concentrated. If the capacity for water reabsorption in the large intestine is not sufficient, the faeces will have a soft to liquid consistency (i.e. osmotic diarrhoea).

The osmotic effect of sugar alcohols caused by excess sugar alcohols in the intestinal cavity can in some cases lead to "osmotic diarrhoea". It must not be confused with pathological diarrhoea, which is infectious and characterised by massive disturbances of water balance and electrolyte loss that can lead to dehydration.

Carbohydrates which reach the large intestine are used as substrate for the micro-organisms in the intestinal flora. The end products of the bacterial fermentation are various gases (notably hydrogen, carbon dioxide, methane) and volatile fatty acids. The more material that reaches the large intestine, the greater the gas production. These gases are partly absorbed and released via the lungs, partly used by the body and partly leave the body in the form of flatulence.

Comparison of the Tolerability of the Various Sugar Alcohols

There is certainly a difference between the different sugar alcohols, and a food technologist designing a sugar-reduced chocolate recipe is well advised to take this aspect into consideration.

A simple comparison of osmotic pressures shows that the osmotic pressure of a monosaccharide sugar alcohol is much higher than disaccharides. Sorbitol, mannitol and xylitol should therefore be avoided for sugar-free chocolate formulations.

There is nevertheless a difference between the disaccharide sugar alcohols. There are exhaustive scientific publications on this topic and threshold figures are quoted, but in actual fact scientifically sound studies which compare the different sugar alcohols for potential gastrointestinal risk are virtually non-existent.

In the Department of Biological Sciences at the University of Salford, there is a team which has researched extensively in this area and a battery of tests have been developed and tested to screen prototype recipes. These include:

In vitro faecal fermentation systems to evaluate the fermentation properties of sugar alcohols.

Breath testing which enables a non-invasive quantification of gases such as hydrogen produced by in vivo fermentation and which provides an indication of potential gastrointestinal intolerance symptoms that may occur[24].

Field study: this is usually a double blind cross-over study with up to 100 participants. Different dietary regimes (including acute and chronic consumption of polyol) have been researched. Laxative symptoms following the consumption of different amounts of a given sugar alcohol are presented in Table 7.

Table 7 Acute Polyol Tolerance Study. Percentage of Subjects in Study Group Passing Loose Stools following Consumption of Polyol-Containing Food

Where:

Mild laxation = The passage of 1 loose stool
Moderate laxation = The passage of 2 loose stools
Severe laxation = The passage of 3 or more loose stools

Dose	No symptoms	Mild	Moderate	Severe
50g Polyol	70.1	15.3	7.1	7.1
35g Polyol	86.2	8.2	4.6	1.0
20g Polyol	95.9	4.1	0	0
0g Polyol	93.9	4.1	2.0	0

Various sugar alcohols have been compared to quantify the potential gastrointestinal symptoms and determine which sugar alcohol has the minimal effect. This research has been entirely sponsored by industry and it is unlikely that the data will be released for publication in the short term. This sort of knowledge creates a competitive edge!

It was surprising to find that the gastrointestinal risk, as evaluated in extensive field studies using chocolate as the test food, does in fact differ considerably from earlier published data which has been used by food legislators to regulate laxative warning declarations on food products. Laxative warnings are obligatory (Table 8).

However, the key to a successful consumer product, as far as gastrointestinal risk is concerned, is the correct choice of sugar alcohol and limitation of the portion size.

Table 8 Polyols can cause intestinal discomfort
Tolerance warnings in 3 countries

Country	Sweetener	Warning
U.K.	Sorbitol, mannitol, xylitol, isomalt, lycasin, lactitol.	When a diabetic claim is made, must state: "best eat less than 25g of X a day". If several substances used: "best eat less than 25g of a combination of X a day".
U.S.A.	Sorbitol if daily intake > 50g Mannitol if daily intake > 20g	Excess consumption may have a laxative effect.
France	Sorbitol, mannitol, xylitol, maltitol, isomalt, lactitol, polydextrose.	Do not give to children under 3 years of age. Excessive daily consumption can cause mild gastrointestinal problems.

10 NON-SUGAR CHOCOLATE PRODUCTS IN THE MARKET PLACE

First Generation - for diabetics

Non-sugar (sucrose) chocolates have been commercialised for many years and positioned as diabetic chocolates. The original products were based on sorbitol to replace the bulk, with saccharin or cyclamate to incorporate the sweetness. Sorbitol is, however, difficult to work with from a chocolate manufacture stand-point; the chocolate has a particular taste and it is also highly laxative.

Sorbitol-based diabetic recipes are increasingly rare now. An example of a German product is presented in Figure 3. The sugar (sucrose) bulk is replaced by a mixture of sorbitol and mannitol, and the intense sweeteners are cyclamate and saccharin. The product is sold in the format of a 100g tablet. Acute consumption of the whole tablet may provoke undesirable gastrointenstinal effects, but, in practice, diabetics have learnt to consume small amounts. Furthermore, diabetics have, over the years, acquired a taste for this type of chocolate and this may be the reason why the manufacturer has not modified the recipe.

Figure 3 Diabetic chocolate made with sorbitol/mannitol and cyclamate/saccharin

Plain varieties contain less than 0.2% total simple sugars but milk chocolate and white chocolate will contain about 9% and 14% of total simple sugars respectively (Table 4). This is due to the sugar (lactose) in the milk ingredient.

Second Generation - modified recipes for diabetics

In these chocolate products the sugar (sucrose) ingredient is replaced by the sugar alcohols isomalt, lactitol or maltitol and with the intense sweeteners aspartame or acesulfame K. As maltitol has a sweet intensity approaching that of sugar, in some recipes the addition of intense sweetener is not necessary. The sugar (sucrose) content will be similar to the previous generation of products but larger portions can be tolerated.

An example of such diabetic chocolate is shown in Figure 4.

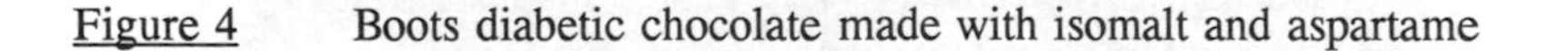

Figure 4 Boots diabetic chocolate made with isomalt and aspartame

Third Generation - "light" in sugar for the mass market

Similar recipes are now positioned for the mass market as LIGHT. A number of products have appeared in France over the last few years. An example has been "Poulain Light". The original product positioning was not entirely clear because there was some initial consumer confusion over the product claim, as the word "light" was often mistaken to mean "light in calories". Only after detailed examination of the wrapper was it clear that "light" referred to sugar. This was then modified and a more recent version of the wrapper clearly stated "light in sugar". The product was available in plain and milk chocolate versions. The portion size was 75g (as compared to 100g which is the standard tablet format) presumably for cost reasons and also to reduce the laxative risk (although the wrapper did carry a laxative warning) (Figure 5). Nevertheless the product positioning has since moved on and a larger range, including both tablets (75g) and countlines (32g) was recently lauched under the name "Poulain Ligne gourmande". The statement "without sugar" is still clearly prominent. The Poulain case is an exciting example of how non-sugar chocolate is evolving in the market place. The bulk sugar replacement ingredients are lactitol and polydextrose, and the intense sweetener is aspartame. This product range has an excellent taste and is now well established in France.

Under current French legislation, such a product for the mass market can be called "chocolate" whereas in the United Kingdom regulations and in the EEC chocolate directive such products containing sugar alcohols and intense sweeteners are called "chocolate flavoured" products. It is thought that the French position will be adopted in the near future.

Figure 5 Poulain "light" made with lactitol, polydextrose and aspartame

Fourth Generation - dental health claims

Boots market a product called "Sugar-free White Chocco Bar" (Figure 6). It is positioned as a chocolate flavour bar with no sucrose. The product does of course contain lactose which is a component of the milk ingredient. A guide to Sugar-free Confectionery and Dental Health was issued to accompany this launch and some experimental data has been included in the brochure to demonstrate the benefit of such a product as compared to a standard sugar-containing product. The "Sugar-free Chocco Bar" is positioned for children and the Boots company have taken a very responsible approach in that portion size is small (30g), limiting the risk of gastrointestinal discomfort. The product distribution is limited to Boots outlets.

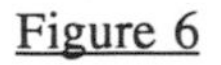
Figure 6 The Boots sugar-free chocolate flavoured bar made with isomalt and aspartame

The first mass market reduced sugar chocolate-flavoured product in the United Kingdom is CHOC KIND by Meltis plc (Figure 7). The sugar sucrose ingredient has been replaced by lactitol and aspartame. The product is presented in bars (33g), in both milk and white chocolate flavours.

Figure 7 The Choc Kind range from Meltis, made with lactitol and aspartame

The following product claims are clearly stated:

- less than 1% sugar sucrose
- with 75% less sugar
- kinder to teeth as part of a sugar reduced diet

The products have an excellent taste and are indistinguishable from standard sugar chocolate.

Fifth Generation - totally sucrose-free and lactose-reduced/lactose-free

These are truly non-cariogenic chocolate-flavoured products and pass the telemetric (tooth-friendly) test.

In addition to the replacement of the sugar ingredient, the milk powder in milk and white chocolate is replaced either by:

- milk protein concentrate (lactose reduced)* + milk fat
  or
- milk protein isolate e.g. sodium caseinate + milk fat

*typical example: 75% protein, 12% lactose and 1% fat

This particular recipe is therefore both sucrose-free and lactose-free. A typical telemetric curve for such non-sugar milk chocolate is shown in Figure 8 (courtesy of Professor Stosser, Medical University of Erfurt Dental School,

Germany) and would qualify as "tooth-friendly" as originally defined by the Swiss Toothfriendly Association.

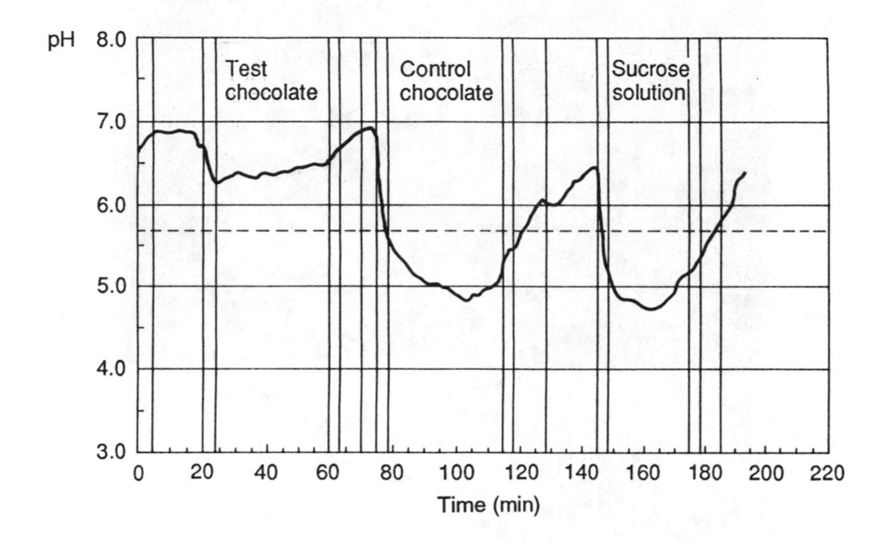

Figure 8 Telemetric Measurement of the pH of Interdental Plaque. The test product is non-sugar (sucrose-free and lactose-free) milk chocolate. The pH in the dental plaque does not fall below pH 5.7 during and for 30 minutes after consumption

An example of a Swiss Toothfriendly chocolate is shown in Figure 9 and the happy tooth symbol is clearly exhibited. The use of sugar alcohol, intense sweetener and lactose-reduced or de-lactosed milk are not currently permitted in the cocoa and chocolate product regulations, neither at the U.K. or EEC level, so such products would need to be described as chocolate flavoured products. The Swiss are clearly setting the pace for non-cariogenic chocolate.

Figure 9 Toothfriendly milk chocolate made with maltitol and lactose-reduced milk powder

REFERENCES

1. NACNE. 'A discussion paper on proposals for nutritional guidelines for health education in Britain'. Health Education Council, London, 1983.
2. W.P.T. James, 'Healthy nutrition: preventing nutrition related disease in Europe', WHO Regional office for Europe, European series; No.24, 1988.
3. Department of Health and Social Security, 'Dietary sugars and human disease'. Report on Health and Social Subjects 37, HMSO, London, 1989.
4. Chocologie, Chocosuisse, Union of Swiss Chocolate Manufacturers Munzgraben 6, 3000 Bern 7 .
5. R. Lees and E.B. Jackson, 'Sugar Confectionery and Chocolate Manufacture', Published by Leonard Hill, Blackie Publishing Group, Glasgow, 1985.
6. B.V. Minifie, 'Chocolate, Cocoa and Confectionery', published by Van Nostrand Reinhold, 1989.
7. The United Kingdom Cocoa and Chocolate Products Regulations (1976), Food and Drugs Composition and Labelling SI No. 541, Amended by SI 1982 No. 17.

8. Council Directive on the approximation of the laws of laws of the Member States relating to cocoa and chocolate products intended for human consumption (1973). 73/241/EEC. Official Journal of the European Communities No. L 228/23.
9. Proposal for a Council Directive on the use of Claims concerning Foodstuffs (1992). Doc. SPA/62/ORIG-Fr/Rev.2.
10. Proposal for a Council Directive on sweeteners for use in Foodstuffs (1992). 92/C 206/02: COM(92) 255 final-SYN 423. Offical Journal of the European Communities No. C 206/3.
11. T. Imfeld and B. Guggenheim. 'The Swiss Association for Toothfriendly Sweets (The Sympadent Association)'. In: Sugarless - the way forward (Edit A.J. Rugg-Gunn) pp197-210, Published by Elsevier Applied Science, London 1991.
12. A. Bär, 'Toothfiendly sweets, the success of a joint industrial/academic public information campaign'. In: Food Ingredients Europe Conference Proceedings. pp 12-19. Expoconsult Publishers, The Netherlands, 1990.
13. T.N. Imfeld, 'Identification of low caries risk dietary components', Karger, Basel, 1983.
14. A.J. Rugg-Gunn, 'Nutrition and dental health', Oxford University Press, 1993.
15. E. Blaser, and A. Zumbé, 'Sugar compositions with molasses-containing fraction of cane sugar as a source of fluoride', United States Patent No. 5, 182, 114, 1993.
16. Crystalline maltitol (maltisorb$^{(R)}$) in manufacture of sugarless chocolate. Obtainable from Roquette Freres SA, 4 rue Patou, F59022 Lille Cedex, France.
17. Lactitol (lacty$^{(R)}$) - a unique reduced calorie sweetener. Obtainable from Purac biochem bv, P.O. Box 21, 4200AA Gorinchem, The Netherlands.
18. Infopac-Isomalt. Obtainable from Palatinit Sussungsmittel GmbH, Wormser Strasse 8, 6719 Obrigheim-Neuoffstein/Pfalz, Germany.
19. G.M. Gee, D. Cooke, G. Wortley, R.H. Greenwood, A. Zumbé and I.T. Johnson. European Journal of Clinical Nutrition, 1991, 45: 561.
20. J.E. Blundell, P.J. Rogers and A.J. Hull, 'Artificial sweeteners and appetite in man'. In: Low-calorie products (Edit. G.G. Birch and M.G. Lindley) pp 147-170, Published by Elsevier Applied Science, London, 1988.
21. B.J. Rolls, 'Sweetness and Satiety'. In: Sweetness (Edit. J. Dobbing) pp 161-173. Published by Springer-Verlag, London, 1986.
22. P.S. Elias, ADI-Wert und Verzehrmengen von Zuckeraustauschstoffen bei Kindern und Erwachsenen. BGA-Schriften 1:19-24, 1988.
23. R. Grossklaus, Intestinale Nebenwirkungen als Wirkungsprinzip der Zuckeraustauschstoffe. BGA-Schriften 1:32-38, 1988.
24. A. Lee, D. Storey and A. Zumbé, Br. J. Nutr. 1993 (in press).
25. A. Zumbé and R.A. Brinkworth, Z. Ernahrungswiss, 1992, 31, 40.

Development and Marketing of Non-sugar Medicines

K. Sugden and I. G. Jolliffe

RECKITT & COLMAN PRODUCTS, DANSOM LANE, KINGSTON UPON HULL, HU8 7DS, UK

1. INTRODUCTION

Over recent years sugar has received a considerable amount of bad press. The sugar debate has increased consumer awareness, particularly of the downsides, and no matter what scientific arguments are put forward relating to sugar and its substitutes there appears to be an unstoppable momentum to the notion that eating too much sugar is bad for you.

The arguments that rage for and against the use of sugars in food manufacture relate to a series of factors, notably to obesity and the subsequent link with the aetiology of a number of diseases, e.g. diabetes, raised blood pressure, hyperlipidaemia, arterial disease and gallstones, and also to the need of individuals, perceived or otherwise, to reduce calorific intake.

For the pharmaceutical industry, the amount of sugar consumed is insignificant compared to that in the food industry and obesity is not the issue. For medicines, the issues relate to a limited extent to diabetes with subsequent warnings on some packs, but more particularly to dental caries which remains prevalent in the western world. Sugar is the most important dietary factor in the cause of dental caries and its presence at plaque-covered tooth surfaces is essential for caries development. The long term use of sugar based medicines in chronically sick children increases the prevalence of tooth decay, a condition that has been referred to as 'Medication Caries'.

The 1989 report[1] of the Committee on Medical Aspects of Food Policy on Dietary Sugars and Human Disease states ...." Dental caries can occur at any age but those at greatest risk are children, adolescents and the elderly........ An increasing number of liquid medicines are available in sugar free formulations.... When medicines are needed, particularly long term, such alternatives should be selected by parents and medical practitioners..... The Panel recommends that Government should seek the means to reduce the use of sugared liquid medicines".

The message to Pharmaceutical Manufacturers, therefore, is quite clear. Liquid Medicines should be made available to the public in sugar free form, particularly those for chronic administration and aimed at the patient groups most at risk, i.e. children, adolescents and the elderly. The argument can be taken further and be applied both to acute dosing and to all oral solid medicines, other than those directly swallowed. Indeed, there is clear evidence that the Medicines Control Agency in the UK are adopting such a policy and recent experience has suggested that they will grant a Product Licence only if the Manufacturer agrees to develop and market sugar free formulations.

THE AVAILABILITY OF SUGAR FREE MEDICINES

THE PHARMACIST

In the UK sugar free medicines are available for a wide range of ailments. A comprehensive list of such medicines is published by the National Pharmaceutical Association, and covers liquid, suspension and dispersible formulations, some chewables, and both Prescription Only and Pharmacy status products[2]. Minor ailments e.g., cough, cold/flu, pain, allergy, indigestion, constipation, oral hygiene and some of the more severe indications, e.g. infection, gastric ulceration, respiratory and central nervous system agents are covered by this list.

The availability of sugar free medicines is wide. A recent study[3] with Pharmacists in the North West of England, however, has demonstrated that although the majority of Pharmacists recognise that sugar containing medicines are a major cause of dental caries in children, they are more likely to recommend for minor ailments such as the common cold, coughs and sore throats the use of a sugared product rather than one that is sugar free. Moreover, the study revealed that of the seven best selling paediatric medicines, only one was sugar free, and of the fourteen most stocked medicines, only two were sugar free.

The results of this study are surprising and disappointing: Patients suffering from minor ailments often use the Pharmacist as the first line for professional advice. Pharmacists, therefore, have the opportunity to improve dental health through product endorsement and recommendation and should be encouraged to do this.

THE CONSUMER

At face value, to remove sugar from a formulation and replace it with an alternative, is a simple task. This, however, assumes that the role of sugar or sugar substitute is simply to impart sweetness to the formulation and ignores the role with regard to all other perceptual characteristics, such as mouthfeel, viscosity, preservation, flavour and after taste. As with foodstuffs, the organoleptic profile of a liquid medicine or chewable tablet and particularly those available through self purchase is paramount for consumer acceptance. Consumers, loyal to a product or brand, are very sensitive to any change in a formulation and often respond to such changes in a negative fashion, assuming that the manufacturer has in some way cheapened the formulation making it less effective, despite all assurances to the contrary that may be provided through either product labelling or advertising. For this reason, and because it is often impossible to exactly match, in sensory terms, sugar and sugar-free formulations, manufacturers often launch the latter as a line extension and not as a direct replacement. This practice particularly applies to those branded products that have a significant consumer loyalty and large market share. Manufacturers are very reluctant to risk the future of such products, even if armed as they invariably are, with preference test data from Market Research to support the case of the sugar free product. Marketing a sugar free variant as a line extension also offers other advantages, allowing for consumer choice and market segmentation.

Despite the awareness amongst the public of the health benefits provided by sugar free formulations, the switch to such variants does not take place overnight but is a gradual process. This has been illustrated from an analogous situation in the Swiss confectionery market, data from which shows a steady growth of the sugar substituted segment from approximately 6% in 1980 to 20% in 1990[4]. The success of this growth lies partly as a result of the work of the Sympadent Association in Switzerland in promoting 'tooth friendly' products, including use of an on-pack logo.

A similar steady growth is experienced with sugar free medicines: Junior Disprol Suspension, the first sugar-free, liquid paediatric analgesic, was launched in 1985. Since that time, a number of other branded medicines have followed and by 1993, sugar free products have 45% of the market segment and are growing annually at the expense of the sugar containing medicines.

Whilst some success of the sugar-free confection market is due to the tooth-friendly, on-pack symbol, this route has not been adopted by the Pharmaceutical Industry, which wishes to avoid the use of logos that may cause confusion between medicines and sweets in the minds of children.

SWEETENING AGENTS

Sugar has a number of useful properties as a pharmaceutical excipient. It is a natural substance, a mainstay food ingredient, in abundant supply and therefore inexpensive. Glucose is a major constituent of honey, an ingredient often utilised in pharmaceutical remedies for minor ailments.

More than this, sugar is also very useful to mask the taste of pharmaceutical active ingredients that may otherwise be organoleptically unacceptable. Sugar provides viscosity and body, essential requirements in liquid formulations. Sugar solutions can be self preservatory, and, moreover, suppress solubility of unpleasant tasting drugs. This property is also important to avoid Ostwald Ripening, a phenomenon that results in crystal growth during product storage.

ALTERNATIVES TO SUGAR

The British National Formulatory describes oral liquid preparations as sugar free if they do not contain fructose, sucrose or glucose. The alternatives to these substances can be divided into two categories, bulk and intense sweeteners.

COMMONLY USED SWEETENERS OTHER THAN SUCROSE

Bulk	**Intense**
Mannitol	Saccharin
Xylitol	Cyclamate
Sorbitol	Aspartame
Hydrogenated Glucose Syrup	Acesulfame

BULK SWEETENERS

To remove sugar from a formulation and replace with another bulking agent is a simple task in practice, but it is very difficult to exactly match the sensory characteristics of sugar and sugar-free formulations. The polyols, mannitol, sorbitol and xylitol are frequently used for this purpose. Directly compressible grades are available, and the crystalline forms impart a smooth mouthfeel to solid formulations. These properties, together with the negative heat of solution for sorbitol, and in particular for xylitol, make them ideal candidates for chewable or lozenge formulations. Their sweetening power, similar or only slightly less than that of sucrose, is sufficient to impart favourable characteristics to the formulation.

Hydrogenated Glucose Syrup has a mild sweet taste, slightly less than that of sucrose. It imparts viscosity and body to liquid formulations and as such has found widespread use.

One drawback of bulk sweeteners is their potential to induce side effects, and they are well known for their flatulant and laxative properties.

LAXATIVE EFFECT OF BULK SWEETENERS

Sweetener	**Laxative Dose (Grams)**
Sorbitol	10 to 50
Xylitol	10 to 50
Mannitol	50
Hydrogenated Glucose Syrup	50
Glycerol	3

The laxative effect is of particular concern. The commonly accepted lower limit for this effect for sorbitol and xylitol is approximately 20 grams, although it has been found with doses as low as 10 grams. There is also a potential issue with glycerol, a sweet non-aqueous solvent that has found application in liquid vitamin formulations. The laxative dose of this substance is particularly low.

The other disadvantage of these bulk sweeteners is their relative cost.

RELATIVE COST OF BULK SWEETENING AGENTS

	Relative Price (per Unit Weight)
Sucrose	1.0
Sorbitol	1.1
Sorbitol DC	2.5
Xylitol	6.8
Xylitol DC	8.5
Mannitol	2.8
Mannitol DC	6.5
Hydrogenated Glucose Syrup	1.4

DC = Direct Compression Grade

All alternatives are more expensive than sucrose and have an adverse effect on production costs. This is particularly so for the direct compression grades, the use of which avoids additional processing steps such as granulation.

INTENSE SWEETENERS

The sensory characteristics of intense sweeteners are given overleaf:

PROPERTIES OF INTENSE SWEETENERS

	Relative Sweetness	Impact	Aftertaste	Instability
Saccharin	400	Slow	Bitter	-
Cyclamate	40	Slow	-	-
Aspartame	200	Clean/Lingers	None	Heat/Low pH
Acesulfame	200	Rapid/Lingers	None	-

Relative Sweetness: Sucrose = 1

The debate in the food industry with regard to the use of sweeteners is intense, to the degree that consumers know them to be additives, covered by E numbers. This adverse publicity particularly surrounds cyclamates and saccharin: cyclamate, although permitted and traditionally used in medicines, is currently banned by legislation in foodstuffs in the UK. This ban will be lifted in the near future when the EC Directive on Sweeteners comes into being, but with the detrimental publicity surrounding the lifting of this prohibition it would be unwise for a Pharmaceutical Manufacturer to use it in a new product. This is particularly so since implementation of the EC Labelling Directive on Medicinal Products, (effective on new products from January 1994) will enforce the declaration on pack of some excipients, including intense sweeteners, knowledge of which is important for safe and effective use. Cyclamate is not permitted in medicines in the USA, other than those with approved new drug applications. This further restricts its appeal, particularly in those products manufactured at a single source and distributed to a number of territories.

Saccharin also suffers from adverse publicity. It does have wide regulatory approval, but has particular problems in the USA where following experiments in rats, the FDA has imposed a mandatory warning label on the pack of all products using this sweetener, viz:

"Use of this product may be hazardous to your health. This product contains saccharin which has been determined to cause cancer in laboratory animals".

There is considerable consumer awareness about the use of saccharin. In the twelve month period between July 1991 and July 1992, Reckitt & Colman received 37 complaints from consumers relating to saccharin in their products. These complaints could be divided into three main categories.

NUMBER OF COMPLAINTS	REASON
24	Bitter aftertaste
7	+ Possible allergy
6	# Safety

+ Although allergenic, the irritance with saccharin is low compared with, for example, the azo dyes.

 # These complaints refer to a USA rodent bladder cancer study in the late 1970s.

Whilst the adverse aspects of intense sweeteners remain clearly in the public eye, the concern should be put into perspective in relation to other additives: UK consumer research undertaken in 1992 (Reckitt & Colman data on file) to test attitudes towards ingredients in analgesic medicines demonstrated that approximately 20% of users were aware of the poor publicity surrounding saccharin. However, more than twice as many users were aware of issues surrounding artificial flavours and artificial colours.

Aspartame and acesulfame are more modern than either cyclamate or saccharin and are finding increasing applications. Aspartame has the drawback of being unstable at low pH, but the breakdown products do not affect taste and off-flavours do not develop.

Furthermore, aspartame has a clean taste reminiscent of sugar and with no after taste. This is in contrast to saccharin, which has a secondary, bitter attribute, readily detected, and which increases with increasing concentration.

Aspartame has wide regulatory approval for use in medicines, although products sold in the USA must carry the pack statement "Phenylketonurics: contains Phenylalanine". The sweet taste of acesulfame is readily detected, persisting longer than that of sucrose and like aspartame it has no after taste at moderate concentrations. Unlike other sweeteners, acesulfame has a pronounced synergistic effect with aspartame, cyclamate, and sorbitol. The international regulatory status of acesulfame is fairly wide, although approval for use in medicines in Australia is still pending.

One of the issues that prevents a faster switch away from the old sweeteners to aspartame and acesulfame is cost:-

RELATIVE COST OF INTENSE SWEETENING AGENTS

	Relative Price (per unit weight)
Sucrose	1.0
Saccharin	5.1
Cyclamate	0.5
Aspartame	12.6
Acesulfame	10.2

Although cyclamate has approximately 20% of the sweetening power of aspartame and acesulfame, and therefore is required in larger quantities per unit dose, it remains very much cheaper, as does saccharin, than the other two substances.

PRODUCT DEVELOPMENT

The use of sweeteners other than sugar to formulate medicines is well established. Direct compressible grades of bulk sweeteners are commercially available and standard formulations for all types of tablet, lozenges and pastille are available. Xylitol and sorbitol, in particular, are very useful in the manufacture of products such as lozenges, imparting to the formulation a hardness, robustness and mouth cooling characteristics.

Hydrogenated glucose syrup is also well established as an alternative to sugar in paediatric liquid analgesic preparations. One of the main product development issues with this type of formulation is an ability to suppress Ostwaldt Ripening. This phenomenon of crystal growth, observed during product storage, is caused by fluctuating ambient temperatures influencing the solubility of excipients. Paracetamol suspensions are particularly prone to this. Sucrose has been demonstrated to inhibit this crystal growth through suppressing the solubility of other materials, to a level where temperature changes have little effect with minimal subsequent crystal nuclei formation of excipients. Hydrogenated Glucose Syrup, although not as effective as sucrose, has been found to control Ostwaldt Ripening to an acceptable degree and has become the excipient of choice in paediatric paracetamol suspensions.

CONCLUSION

It is widely accepted that sugar containing medicines are a cause of dental caries. There are now a large number of sugar free formulations commercially available to treat a range of indications, and whilst the use of these alternatives is steadily growing, there remains a role for healthcare professionals to heighten consumer awareness.

Pharmacists, in the first line for the self-purchase of medicines to treat minor ailments, are ideally placed to recommend the appropriate medication to the consumer.

Issues remain about the application and safety of the alternatives to sugar, particularly the intense sweeteners, and cost of goods also remains a problem for the manufacturers.

REFERENCES

1. Department of Health Report on Health & Social Subjects, 37: Dietary Sugars and Human Disease. Committee on Medical Aspects of Food Policy. HMSO, London, 1989.

2. National Pharmaceutical Association, London. Information Leaflet, Sugar Free Medicines. 1992.

3. I C Mackie, H V Worthington and P Hobson. Pharm J. 1992, May 9th 621.

4. T Imfeld and B Guggenheim in: Sugarless, The Way Forward. Ed. A J Rugg-Gunn, Elsevier, London 1991.

Sugar-free Gum – A Success Story

Adrian Piotrowski

DENTAL PROGRAMMES MANAGER, THE WRIGLEY COMPANY LIMITED, ESTOVER, PLYMOUTH, DEVON PL6 7PR, UK

1. INTRODUCTION

The purpose of this article is to demonstrate the success of sugar-free gum in the U.K., and to provide an insight into why the category has been so successful.

The article is broken down into three sections. Firstly, a brief overview of the U.K. Confectionery Market, its size, the growth in the market, and its composition. Then an overview of the U.K. Chewing Gum Market, looking at the history of the category, its size, sector shares etc. Finally, a focus on sugar-free gum, particularly highlighting the key movements and reasons behind the growth of this sector over recent years.

2. U.K. CONFECTIONERY MARKET

The U.K. Confectionery Market is probably one of the most sophisticated and developed confectionery markets in Western Europe.

We consume about 810,000 tonnes of confectionery per year in the U.K., and spend some £4 billion on confectionery. Relative to other markets, it is certainly amongst the largest.

To put that into perspective - as a nation we spend around £12.8 billion on beer, £9.9 billion on tobacco, £4.6 billion on wine. Then comes confectionery, frozen foods at £3.7 billion, carbonated soft drinks at £3.5 billion and newspapers and magazines at £3.3 billion. Things like coffee and tea are relatively small markets in relation to these larger markets at around £600 to £700 million.

Figure 1 shows how the confectionery market has fared over the last 25 years, highlighting growth from around 606,000 tonnes in 1967, to 810,000 tonnes in 1992. The per capita consumption, also included in Figure 1, has remained fairly flat, however, particularly over the

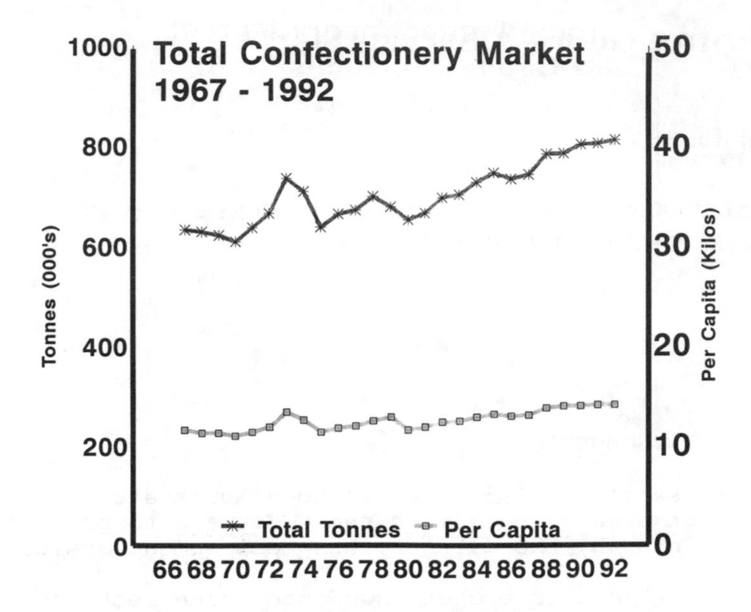

Figure 1 The U.K. Confectionery Market Tonnage and Per Capita Consumption, 1967 - 1992

last five years. In 1967 each person in the country was consuming around 12.0 kilos of confectionery, i.e. 34 grammes per day. Today the figure is closer to 14 kilos, i.e. 38 grammes per day, a little more than half a Mars Bar. But looking at the trend between 1988 and 1992, that flat trend can be seen, with per capita consumption changing very little over the period.

One sees a slightly different picture in Figure 2 when looking at what it is that people are consuming, with chocolate increasing fairly significantly over the period, from around 6.5 kilos per head to 8.8 kilos, while sugar confectionery has remained flat at around 5.2 kilos per head.

That trend is of course reflected in the overall market trend, where, between 1967 and 1992, chocolate confectionery increased from 332,000 tonnes up to 482,000 tonnes, while sugar confectionery declined from a high of 348,000 tonnes in 1973 down to 299,000 tonnes in 1992.

Looking at the composition of the chocolate market, it is the filled bars, blocks and countlines, and boxed chocolate sectors which have been responsible for most of the growth over the last twenty five years, with filled bars, blocks and countlines, e.g. Mars, Topic, etc., accounting for almost half of the markets.

Boxed chocolates have made something of a comeback, increasing share from 18% to 25% of the market.

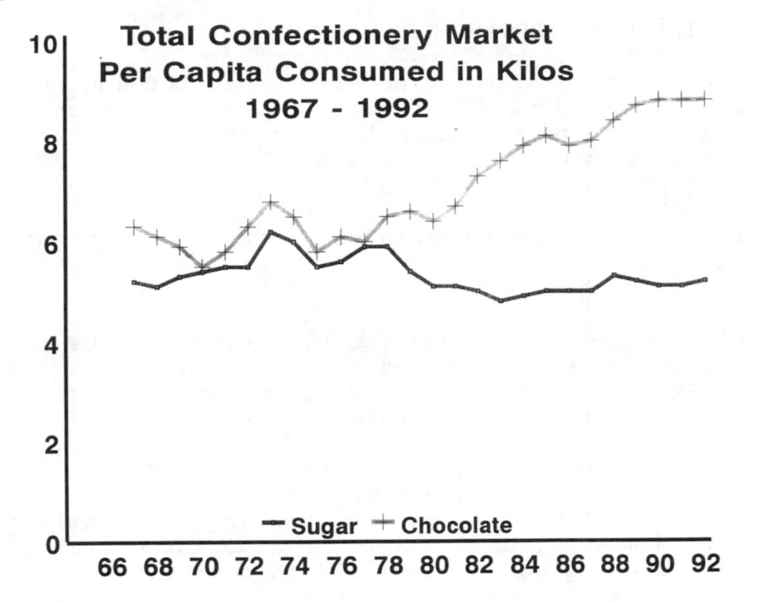

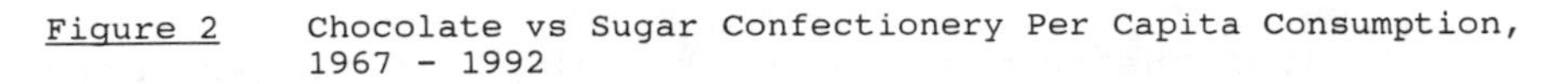
Figure 2 Chocolate vs Sugar Confectionery Per Capita Consumption, 1967 - 1992

The sugar confectionery market meanwhile, has very much moved away from being a market of boiled sweets and toffees, which accounted for around 55% of the market in 1967, to one where things like mints, licorice, nougat, popcorn etc. have expanded from 25% of the market in 1967 to over 42% today. Chewing gum has grown from 4% to 6% of the sugar confectionery market over the same period.

The Confectionery Market is very much dominated by Nestle, Cadburys and Mars, who collectively account for around 60% of confectionery sales in the U.K. As you would expect, it is very much their brands that account for the top selling confectionery lines in the U.K. Kit Kat is worth over £200 million, Mars Bar £135 million, Twix and Cadbury's Dairy Milk in excess of £90 million, Roses, Quality Street and Snickers in excess of £80 million, Galaxy, Maltesers, Aero, Smartees and Bounty in excess of £40 million and Orbit sugar-free comes in at around £34 million for 1992.

Summarising this first section on the U.K. Confectionery Market, the total market is worth around £4 billion, and comprises some 810,000 tonnes. Chocolate confectionery accounts for the biggest proportion of the market, with around two thirds of total sales. Per capita consumption has remained generally flat over the last 5 years, at around 14 kilos per head. And chewing gum is a relatively small player in the field, accounting for 6% of sugar confectionery.

3. THE U.K. CHEWING GUM MARKET

In spite of its relatively modern image, chewing gum has in fact been on the scene for centuries. For example, the ancient Greeks are known to have chewed a Mastic gum, which in fact was a resin from the Mastic Tree. Grecian women, in particular, favoured this gum, and used it for cleaning their teeth and sweetening their breath. The Indians of New England were also known to be big chewers of resin from the Spruce Tree and the American settlers picked up the chewing habit from them.

Modern chewing gum had its beginnings in the 1860's when Chicle was first brought to the U.S., and it wasn't until 1892 that William Wrigley Jr. began manufacturing chewing gum in the United States.

Wrigley's chewing gum was introduced to the U.K. in 1909, and from small beginnings, the U.K. operation now employs over 600 people and produces around 25 million pieces of gum every day. While on a worldwide basis, Wrigley's chewing gum is now sold in over 100 countries. The Company has a turnover of over $1.2 billion, and employs some 6,500 people.

In 1992 the chewing gum market in the U.K. was worth around £112 million at retail selling prices, with around 800 million packets of gum sold. As previously stated, relative to the confectionery market as a whole, the chewing gum market is still very small, accounting for just 6% of sugar confectionery sales. But over the past 25 years, the market has grown from £9.3 million to £112 million, and in fact, has more than doubled in the past six years alone (Figure 3).

Within the market we've seen some significant changes in the chewing gum sectors. Stick or slab gum has become more popular, and now accounts for 65% of the market, versus 59% in 1967. But by far the most dramatic movement has come from sugar-free gum, which wasn't even introduced until the 1970's, but in 1992 accounted for over half of the sales in the gum market.

Wrigleys hold the strongest share of the gum market, accounting for some 87% of chewing gum sales, and it is the Company's sugar-free brands, Orbit and Wrigley's Extra, which now lead the market, holding first and second positions with a combined sterling share of 56%.

Around one third of the population chew gum. The habit is most prevalent amongst 11 - 24 year olds. There is a slight female bias, and the habit tends to be most prevalent amongst the C1, C2 socio-economic groupings. Having said that, the market has seen a dramatic influx of new chewers over the past two years, which has come from all sectors of the population. Indeed, in the second half of 1992 some 4.5 million new chewers entered

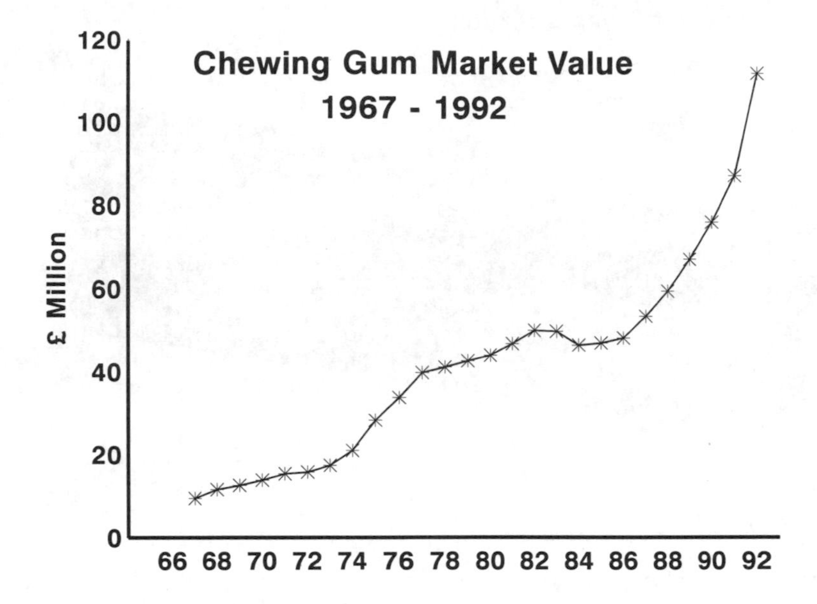

Figure 3 U.K. Chewing Gum Market Value, 1967 - 1992

the market.

In summary, the chewing gum market in 1992 was worth around £112 million, having increased in value by over 100% since 1987. Wrigley has the largest share of the market with around nine out of ten packets bought being Wrigley's chewing gum. The most dramatic movement in the market over the past six years has been that towards sugar-free gum, to the point today where sugar-free gum actually outsells the traditional sugar gums.

4. SUGAR-FREE CHEWING GUM

This final section addresses why it is that the sugar-free chewing gum has not only managed to compete with the sugar chewing gum brands, but has actually outstripped them.

Sugar-free gum as we know it today, ie. gum substituting sugar for sugar alcohols or alditols and high intensity sweeteners, was first commercially made and distributed in Denmark in the mid 1960's.

The first sugar-free gums introduced in the U.K. were 'Trident', and a product called 'Skels', which were both introduced in the mid-1970's. These were quickly followed by the introduction of Wrigley's Orbit in 1976 (Figure 4). Both Trident and Orbit were introduced on a tooth care platform, i.e. avoidance of sugar.

At that time the products were, without question,

Figure 4 Orbit, introduced in 1976

inferior to their sugar-containing gum equivalents, and suffered from textural and short shelf-life problems, which meant that they very quickly became brittle, crumbly and stale.

In spite of this product inferiority, the sugar-free sector grew to around 10% of the market by 1985, very much fuelled by the general consumer trend towards a more healthy lifestyle, something which was reflected in the T.V. advertising for Orbit at that time.

The principal polyols used in sugar-free gum up to 1985 were Sorbitol, Maltitol, Mannitol and Maltitol Syrup, with Saccharin being the most commonly used intense sweetener at that time.

Because of the cost of bulk sweeteners relative to the cost of sugar, (Sorbitol is around 2 to 3 times more expensive than sugar and Xylitol is even more expensive), sugar-free gum has always sold at a premium stick-for-stick or pellet-for-pellet versus the sugar equivalent. One might have thought therefore that a 10% share of the market was as high a share as sugar-free gum would ever take.

1986 however, was to see a dramatic acceleration of growth in the sector, very much fuelled by the reformulation of Orbit with the intense sweetener Aspartame, better known to consumers as NutraSweet. Wrigley's had also addressed the textural problems of the original formula within the new formulation, and had perfected the technique of encapsulating the intense sweeteners to ensure their release over an extended chew period, thus giving consumers long lasting taste delivery. Consumer tests prior to launch confirmed a 2:1 preference for the new formula over the old one, and this was quickly realised in sales which increased two fold in the period between 1986 and 1989. All of this in spite of the fact that the formulation had no direct advertising support, but rather limited promotion, confined to an on-pack message, and support on the gum display unit.

1990 saw a further acceleration of sugar-free chewing gum volume, as Wrigley's introduced a sugar-free pellet gum called Wrigley's Extra (Figure 5) to cater for those consumers who preferred the pellet format. Again, the product contained the intense sweetener, Aspartame, but this time a blend of Xylitol and Sorbitol was used to provide a different taste perception to that of Orbit, and to capitalise on the cooling qualities of Xylitol. Packed in a ten pellet pack, and again, retailing at a premium to the sugar equivalents, Wrigley's Extra was initially launched in a Peppermint flavour, and quickly established a strong foothold in the market.

The ranges of both Wrigley's Extra and Orbit were later extended with fruit variants, thus extending the consumer choice of sugar-free flavours out of purely mint variants.

By 1991, sugar-free chewing gum was accounting for around 38% of all chewing gum sales, but the most dramatic growth was still to come.

5. CHEWING GUM AND DENTAL CARE

Dental research into the effects of stimulating saliva through the chewing of gum had been going on for some time, and development of interproximal wire telemetric appliances, capable of measuring the acidity of plaque, provided the conclusive breakthrough confirming gum's role in stimulating saliva to neutralise plaque acid fast.

It is known that many foods have a high acidogenic potential. It is also generally accepted that teeth are under attack and the enamel can demineralise when the acid level falls to below pH 5.7. For example, raisins can induce a pH drop to pH 4.03, crisps to pH 4.11 and even apples to a pH of 4.19.

Within minutes of eating, the bacteria in the mouth uses the fermentable carbohydrate to produce harmful acid in plaque. This acid attacks the enamel of teeth which can lead to the formation of cavities.

Nature's solution is to produce saliva to neutralise

<u>Figure 5</u> Wrigley's Extra, introduced in 1990

the acid, but numerous studies have shown that it can take anything up to two hours before the risk to teeth is removed.

Constant snacking, in today's diet, can result in a permanent acid attack on teeth throughout the day.

Plaque on the accessible surface of teeth can of course be reduced or removed by brushing. But brushing after a meal or snack is not always convenient, and plaque in the areas between the teeth is difficult to reach and is generally a high risk cavity area.

It was clear then that sugar-free gum could provide a useful adjunct to a good dental care routine, not through any particular ingredient in the gum but, rather, through its ability to stimulate saliva.

I think it fair to say that some within the dental profession have underestimated the important role of saliva. Others, however, know what a vital role it plays in dental care and some even refer to it as 'the miracle in the mouth'.

So what are the main functions of saliva? I believe the following are key:

- the maintenance of oral pH, especially in the plaque by neutralisation of excess acid

- the removal of food debris and cleansing
- the inhibition of dental demineralisation by limiting acid attack
- the provision of patient comfort by moistening and protecting the hard and soft tissues in the mouth
- the delivery of minerals eg. calcium and phosphate, to create a pH environment in which minor dental damage can be repaired or remineralised
- and finally, the regulation of the oral ecological balance

The chewing of gum stimulates this 'miraculous fluid' by as much as ten fold, and the results of that are fast neutralisation of plaque acid, relief from dry mouth and assistance in the repair of early decay damage, i.e. enhanced remineralisation.

Wrigley decided to focus their communication on chewing gum after meals and snacks to stimulate saliva and neutralise plaque acid fast and used the Stephan curve to demonstrate the difference that gum could make to pH levels after eating. (Figure 6).

The temptation back in 1988 was to dash out and tell the consumer about the dental benefits of chewing gum. Wrigley held back, however, feeling that there was no point in educating the general public, if the profession did not accept the dental benefits. We set ourselves a target therefore of 80% acceptance by the dental profession before we would launch our consumer campaign, and embarked on a very specific communication route through to the consumer.

The first people approached were the Opinion Formers within the dental profession - people involved in teaching and research, and within the professional bodies. They were aware of the research results and happy for us to detail the benefits of chewing sugar-free gum to the dental profession. Over the months and years they acted as invaluable advisors for our dental messages and materials. Only when they were happy would Wrigleys despatch a mailer to dentists, place an advertisement in dental journals or place patient materials in surgeries. And only when the profession were accepting our communication messages were we happy to launch our consumer campaign.

Since 1988 Wrigleys have had a full and varied communication programme to dentists and pharmacists. A programme which has included: professional mailings, journal advertisements, professional conferences/exhibitions, patient literature, patient samples, sugar-free chewing gum in sales displays through dental surgeries and a slide and video programme for schools.

Wrigleys also sponsored the development of a dental

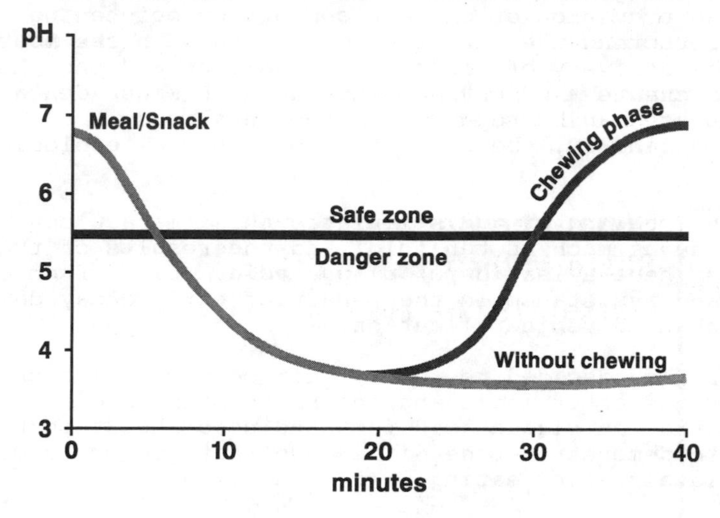

Figure 6 Typical pH response after a meal or snack with and without sugar-free chewing gum

schools teaching kit, enabling dental students to experiment with salivary stimulation and observe its buffering effects. The kit was developed by Professor Geddes at Glasgow Dental School, and its experiments now form part of the curriculum followed by dental students.

By 1990 we had surpassed our target of 80% acceptance by the profession of chewing gum's role in dental health. Indeed a survey of dentists and hygienists in 1991 showed that 93% accepted that sugar-free gum should be part of an effective dental care routine, with 76% actively recommending sugar-free chewing gum to their patients.

Wrigley began to develop T.V. advertising at this time and, in conjunction with public relations activity, launched this in the North West of England in March 1991. The campaign was gradually rolled-out until it was fully national in January 1992. Since then a second T.V. execution has been developed and this is currently being rolled-out across the country.

The T.V. advertising has been further supported by a major sampling campaign which has seen 13 million packets of Orbit or Wrigley's Extra sugar-free gum delivered to households around Great Britain.

The results have transformed the market.

Sugar-free chewing gum sales have increased by 190

million packets in 2 years.

An estimated 4½ million new consumers started chewing gum (almost all of it sugar-free) in the second half of 1992 alone.

And sugar-free gum has increased its share of the market from around 30% at the beginning of 1991 to 56% today (Figure 7).

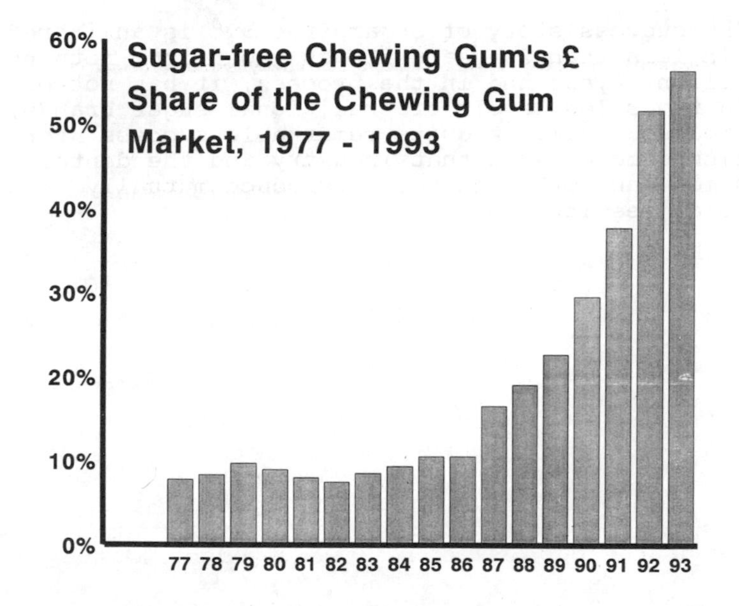

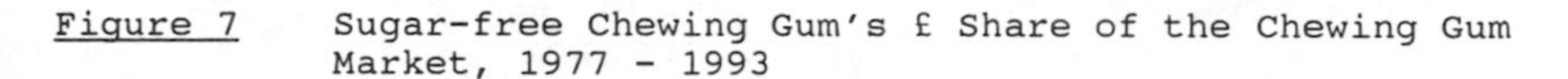

Figure 7 Sugar-free Chewing Gum's £ Share of the Chewing Gum Market, 1977 - 1993

Summarising this third section, following its introduction in the mid '70's, sugar-free gum sales increased to the point where sugar-free brands held around a 10% market share in the early 80's. A superior, reformulated Orbit product in 1986 and the introduction of Wrigley's Extra with Xylitol in 1990 pushed its share up to around 30% of the market by early 1991. Communication of the dental benefits of chewing sugar-free gum to stimulate saliva and neutralise plaque acid, have since pushed annual sales up to £69 million (M.A.T. June '93) and its market share up to 56% - making it larger than all the traditional sugar chewing gum brands put together!

6. CONCLUSIONS

In conclusion, chewing gum is a relatively small player in the £4 billion confectionery market. Total confectionery per capita consumption is around 14 kilos per head.

The chewing gum market itself has doubled in value since 1987 and was worth £112 million in 1992. Wrigley brands account for 9 out of 10 purchases.

Almost all the increase in the market has come from sugar-free gum - through reformulation and new products, and more recently as a result of the dental benefits of acid neutralisation and remineralisation through its ability to dramatically stimulate salivary flow.

The success story of sugar-free gum is an incredible one. In less than 20 years it has grown from nothing to £69 million a year and in the process, it has not only matched the sales levels of traditional sugar brands, but has surpassed them. A quite remarkable success story and one which demonstrates that industry and the dental profession can work together to produce mutually beneficial results.

Subject Index